T0282414

Understanding PCR

Understanding PCR
A Practical Bench-Top Guide

Sarah Maddocks
Cardiff Metropolitan University, United Kingdom

Rowena Jenkins
Cardiff Metropolitan University, United Kingdom

ELSEVIER

AMSTERDAM • BOSTON • HEIDELBERG • LONDON
NEW YORK • OXFORD • PARIS • SAN DIEGO
SAN FRANCISCO • SINGAPORE • SYDNEY • TOKYO
Academic Press is an imprint of Elsevier

Academic Press is an imprint of Elsevier
125 London Wall, London EC2Y 5AS, United Kingdom
525 B Street, Suite 1800, San Diego, CA 92101-4495, United States
50 Hampshire Street, 5th Floor, Cambridge, MA 02139, United States
The Boulevard, Langford Lane, Kidlington, Oxford OX5 1GB, United Kingdom

Notices
Knowledge and best practice in this field are constantly changing. As new research and experience broaden our understanding, changes in research methods, professional practices, or medical treatment may become necessary.

Practitioners and researchers may always rely on their own experience and knowledge in evaluating and using any information, methods, compounds, or experiments described herein. In using such information or methods they should be mindful of their own safety and the safety of others, including parties for whom they have a professional responsibility.

To the fullest extent of the law, neither the Publisher nor the authors, contributors, or editors, assume any liability for any injury and/or damage to persons or property as a matter of products liability, negligence or otherwise, or from any use or operation of any methods, products, instructions, or ideas contained in the material herein.

Library of Congress Cataloging-in-Publication Data
A catalog record for this book is available from the Library of Congress

British Library Cataloguing-in-Publication Data
A catalogue record for this book is available from the British Library

ISBN: 978-0-12-802683-0

For information on all Academic Press publications
visit our website at https://www.elsevier.com

Working together
to grow libraries in
developing countries

www.elsevier.com • www.bookaid.org

Publisher: Sara Tenney
Acquisition Editor: Jill Leonard
Editorial Project Manager: Fenton Coulthurst
Production Project Manager: Lucía Pérez
Designer: Mark Rogers

Typeset by TNQ Books and Journals

Contents

Author Biographies

Sarah E. Maddocks is a lecturer of Microbiology at Cardiff Metropolitan University, the United Kingdom. She received a PhD from the University of Reading in 2005 and worked on bacterial virulence before embarking upon investigations into the host—pathogen relationship during infection. Her main research interests involve the study of wound colonization, bacterial persistence, and the shift from colonization to infection.

Rowena Jenkins is a lecturer in Microbiology at Cardiff Metropolitan University, from which she received her MSc in Biomedical Science and Medical Microbiology and PhD, and a fellow of the Higher Education Academy. Her main research has focused on the antimicrobial effects of Manuka honey on potentially pathogenic microorganisms in health care and wound management. Cellular morphology, physiology, biofilm prevention/disruption, adhesion/invasion, virulence expression, and proteomic/genetic expression profiles of organisms such as MRSA and *Pseudomonas aeruginosa* have been studied in response to varying treatments with honey. The data generated by these in vitro studies help reveal how antimicrobial agents might be applied in clinical settings. Rowena is also exploring how novel antimicrobial agents can potentiate antibiotics already in use but whose effectiveness is now challenged by antimicrobial resistance.

Chapter 1

Things to Consider Before Preparing to Undertake Your First Polymerase Chain Reaction

ABSTRACT

This chapter provides practical advice on what needs to be addressed before undertaking polymerase chain reaction (PCR). From keeping the workspace nuclease free to recipes and shopping lists; information that is vital know and understand before putting on a lab coat and entering the lab, is described. The reader is guided through the practical aspects of preparing to undertake PCR with examples of reagents provided as well as a list of common manufacturers of PCR reagents. Safety considerations are discussed and thought is given to equipment and reagents that will be required for analysis of the first PCR reaction.

1.1 INTRODUCTION

So, you have decided to tackle the polymerase chain reaction (PCR); this is an extremely powerful technique that will enable you to amplify fragments of DNA. PCR is also incredibly versatile, and once you have mastered the basics, you will be able to apply the techniques you have learned to a host of molecular techniques including genetic profiling, detection, gene expression, and genetic modification. The world of PCR can be overwhelming, from deciding on your primer design, to where to buy your polymerase. It is therefore important that you understand the fundamental requirements necessary to prepare and carry out PCR, to avoid months of frustration or results you can not even begin to comprehend.

There are a number of factors to consider when planning to undertake PCR for the first time, and if these are addressed prior to putting on your lab coat and picking up your pipette, then the whole process will run much more smoothly, meaning you are more likely to be successful. Such considerations include aspects of safety, preparation of reagents, and consideration of reaction parameters and organization of your work area. This chapter will describe some of the considerations you need to address before heading into the lab; it is a good idea to equip yourself with a pen whilst reading through this chapter so that you can begin to plan your first PCR.

1.2 BASIC LABORATORY PREPARATION

Saying that it is important to work in a clean, organized, and orderly manner might seem obvious but is imperative for successful PCR. PCR is a powerful procedure in which small quantities of DNA are amplified; the success of PCR can be affected by the presence of contaminants from the environment or due to poor handling of reagents, or PCR inhibitors which are often inherently present as a consequence of DNA preparation. Working in an area that is free from possible sources of contamination or potential inhibitors is therefore paramount. So first, ensure that your bench is free from clutter, including left over experiments, reagents, used gloves, or plastics, to name a few, that could contaminate your reaction. It is usually good practice to wipe over your bench with a standard disinfectant or 70% ethanol before you begin—you could also wipe the barrel of the pipettes you will use. Often people use filter tips for PCR to prevent anything that might be in the barrel of the pipette from being inadvertently introduced into the reaction. Sometimes, people use a dedicated "PCR" set of pipettes, but this is not paramount providing you are confident that yours are clean! You will be handling very small amounts of often colorless liquid, so make sure that your pipettes are calibrated to ensure precision. Most of your PCR reagents will need to be kept on ice, so have an ice bucket to hand and also somewhere to dispose of waste, such as used tips and tubes. It goes without saying that you will need to wear a lab coat and gloves (the latter to avoid contamination of your work by you and to protect you from harmful chemicals)—if you think you have contaminated the gloves you are wearing and that you might transfer this to your PCR, then change your gloves. Below is a photograph showing a bench that is setup for PCR (Fig. 1.1). You will notice the sensible work flow, bringing reagents in from the left, preparing the reaction in the central area of the bench, and disposing of used tips or empty tubes into an appropriate jar which is in easy reaching distance.

To ensure that you remember to add each reagent, it is usually a good idea to write yourself a list (an example list of ingredients with hints and tips are given below; volumes will be addressed later) of the ingredients you will add to your PCR reaction and tick them off as you add them; this is also useful when purchasing reagents so that you do not forget anything.

You will usually be setting up more than one reaction at a time if you are, for example, testing a new set of primers, titrating the template material, or preparing replicate reactions. In this case, it is usually a good idea to prepare a master mix of ingredients in one 1.5 mL tube, i.e., for five reactions multiply the volume of each reagent to add by 5, and then divide this up between the

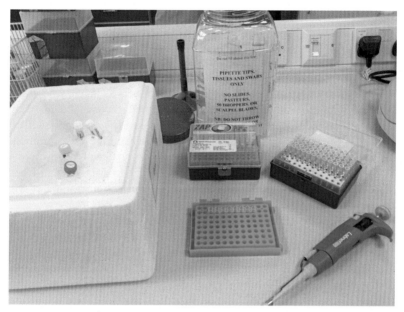

FIGURE 1.1   A PCR workbench.

PCR tubes that you have. This avoids inaccuracies during pipetting, or problems if reagents have not properly mixed.

When using the PCR recipe list, you will most likely still be planning your PCR "on paper," designing primers and reaction parameters—if you are using a predesigned set of primers with known reaction parameters you can move onto the wet work and skip to Chapter 3 which describes how to analyze the results of the PCR reaction.

PCR Recipe List, Hints, and Tips

- Template DNA: Decide how you will obtain this; will you use a commercially available kit or an "in-house" extraction? Check the concentration of the template material using either by UV spectroscopy or by electrophoresis with a quantitative molecular weight marker (see Chapter 3).
- Primers (forward and reverse): If you need to design primers, please turn to Chapter 2 for guidance. A good start point for PCR is to begin by adding 100 pmol per reaction, but you can prepare a series of dilutions to find the optimum concentration to work with. Primers usually arrive lyophilized, and the company you purchased them from will tell you how to reconstitute them (see Chapter 2). Once you have reconstituted your primers, make a working stock of 100 pmol/mL, and then you would not be constantly dipping into you "master" stock).

- dNTPs: These are usually purchased individually or as a premixed preparation containing equal quantities of all four bases. These are usually available at stock concentration of 100 mM (but please check the information from the manufacturer before use) which will be diluted upon addition to the reaction.
- MgCl$_2$/MgSO$_4$: It is vital for PCR to work; it acts as an enzyme cofactor and also impacts the specificity and stringency of primer annealing. Buffers that are provided with commercially available PCR enzymes will often include Mg^{2+} at an amount that will give a working concentration of 1.5 mM when diluted in the reaction. Some enzyme systems provide Mg^{2+}-free buffers and a stock solution of MgCl$_2$ or MgSO$_4$ that is added accordingly, this allows the user to prepare a concentration range of Mg^{2+} as part of the optimization process.
- Reaction buffer: This is usually supplied with the enzyme as a 5× or 10× concentration that is diluted as you prepare the reaction; the reaction buffer may or may not contain Mg^{2+} (see above).
- Nuclease-free water: Sometimes, this is provided with the enzyme but if not can be purchased from a large number of suppliers or can be prepared using nanopure water (see Section 1.3). It is advisable to keep nuclease-free water in small aliquots to minimize problems associated with contamination.
- Polymerase: There is a number of different polymerase available, the most common is Taq. Polymerases with a number of different characteristics are available and so that can be selected according to application. For example, it is possible to buy "long polymerase" or "short polymerase" which as the names suggest are optimized for amplification of short or long sequences. Some enzymes have proof-reading activity and are good for cloning where errors need to be minimized (see Chapter 2).

Once you have considered all of the above parameters, have designed your primers and prepared your template material, it is a good idea to write a reaction-specific "ingredient list" to pin up at your workstation so that you can easily refer to it when preparing your reaction.

1.3 MAINTAINING A NUCLEASE-FREE ENVIRONMENT

Nucleases are the nemesis of any successful PCR. These are found just about everywhere and will rapidly degrade DNA, including primers, template DNA, and your PCR product. PCR reagents are provided "nuclease free," and it is possible to buy additional nuclease-free reagents from most laboratory suppliers—you will find it useful to maintain a good stock of nuclease-free water. It is a good idea to keep reagents in small quantities so that if they should become contaminated, you can throw them away and start afresh. If your budget does not stretch to cover the costs of buying nuclease-free water, you can always prepare your own by autoclaving nanopure water. Again, it is advisable to prepare small aliquots of between 10 mL and 20 mL for this purpose. It is always the best to purchase things like reaction buffers for enzymes and magnesium solutions, and most of these will come with the enzyme when you buy it (see Section 1.6 for where to buy reagents and supplies).

As mentioned previously, if possible, it is a good idea to use filter tips when setting up PCR reactions to help you to maintain a nuclease-free environment; these prevent aerosols associated with your pipettes from getting into your PCR reaction and therefore contaminating it. These can be purchased from a wide range of suppliers. DNase-free and RNAse-free PCR tubes can also be purchased from numerous suppliers (see Section 1.6), these usually arrive in large bags so it is a good idea to portion them out into smaller, sterile containers (if you prefer, you can autoclave them as you would any other microcentrifuge tube). Because you can introduce nucleases to your working environment, it is important that you always wear a lab coat and gloves when handling any PCR reagents, more specifically ensure you wear a "fresh" pair of gloves when beginning PCR work and that your lab coat is not filthy!

Unless you have a dedicated bench space that is only used for molecular work, it is wise to clean your bench space before setting up your reaction; you can use disinfectant or 70% ethanol to do this, but there are some commercially available sprays that are specifically designed to remove nucleases and can be used to surface treat benches and equipment such as pipettes or PCR tube racks. Below is a recipe to prepare a solution that will do this job; if you choose to make this solution ensure that it is prepared using nanopure water and in a glass container, then sterilized by autoclave for 15 min at 121°C prior to use. Treat it as you would any other nuclease-free reagent and prepare only small "working" volumes. If at any point you think you might have contaminated your work space or equipment, it is advisable to clean everything again as there is nothing more frustrating than running a PCR to find later that there are problems associated with the presence of nucleases or other contaminants.

Recipe for Surface Cleaner

Prepare 50 mL 10 M NaOH, and 5 mL 10% (w/v) SDS, add these to 445 mL nanopure water, and autoclave for 15 min at 121°C. NaOH and SDS should be handled wearing gloves; SDS should be weighed out in a well-ventilated area, or wearing a face mask, as it is a respiratory irritant.

1.4 CONSIDERATION OF TIMESCALES

Generally speaking, when considering the time scale needed to set up your PCR, it is best to overestimate the amount of time you think you will need. There is nothing worse than running out of time when trying to prepare the reaction or rushing and making mistakes. If you are using predesigned primers and pre-optimized reaction parameters you can head straight into the lab and get started. However, if this is not the case then there are a few more things to consider.

Depending on the gene you are investigating, you may need to find and identify the gene sequence and design primer pairs (Chapter 2). If you are using novel or newly designed primers, they may not work first time—the concentration, thermal cycling conditions, or other parameters might need optimizing (see Chapter 2 for common problems and troubleshooting). You might find cross-priming if you are looking for more than one gene at a time or the primers you have designed may dimerize (Chapter 2).

You will need to prepare template DNA; this can be done using a commercial kit or you can use your own "in-house" methods (Section 1.4). Either way, this will take time, 30 min to a couple of hours with a commercial kit (depending on the kit you use), or half to a full day, if using your own method. After extracting the DNA, you will have to check its concentration and purity (Chapter 3), and if there are problems at this stage, you might need to repeat the extraction. Each of these adjustments could take a day or more of your time so it is easy to see how, if everything does not run perfectly first time, things might take longer than expected.

Something that people tend to forget to factor in is the delivery time for the reagents. Depending on your ordering system and the supplier, sometimes reagents can take a couple of weeks to arrive; it is always best to ask the company when ordering to see if they can tell you the expected delivery time. However, if you have everything in the lab and your template DNA was successfully extracted, you should be able to perform a standard PCR and gel analysis within a day.

1.5 GENERAL PROTOCOL FOR EXTRACTING DNA

This protocol is good for extracting DNA from eukaryotic cells that do not have a cell wall, it can be adapted to extract DNA from any type of cell that has a cell wall by the addition of specific enzymes (such as lysozyme for bacteria) or by lysis using bead-beating. It is intended to be carried out on a small scale using 1.5 mL microcentrifuge tubes. For lysing bacteria, can either purchase a beadbeater or use a series of vortexing steps, as described here.

To lyse by vortex, first pellet the cells by centrifugation, discard the supernatant and resuspend the pellet in lysis buffer, to this add a small amount of 0.1 mm glass beads (to a depth of 50−100 μL)—you will need to sterilize them by autoclave first—then vortex for 30 s, rest on ice for 30 s, vortex for 30 s 5−6 times and continue as per the protocol below.

Add 100 μL of lysis buffer to your sample (if you are not lysing by beatbeating and/or vortex the sample is simply pelleted by centrifugation, and the supernatant removed prior to addition of the lysis buffer; if you have mechanically lysed the sample, do not add more lysis buffer) and resuspend the material by pipetting.

Lysis Buffer Recipe (Prepare in PCR-Grade, Nuclease-Free Water)

50 mM Tris—HCl (pH 8.5)

1 mM EDTA

0.5% SDS

Store at room temperature

Add 3 μL of a 1 mg/mL solution of RNase A to your sample and incubate the sample for 10 min at 37°C

Next, add 200 μg proteinase K and incubate for a further 10 min at 37°C

You will need to heat inactivate the proteinase K by placing the samples at 95°C for 5 min

Next, cool the samples on ice for 5 min, centrifuge the samples using a microcentrifuge, at maximum speed (usually 13 k rpm) for 5 min

Transfer the supernatant to a fresh microcentrifuge tube

Add one volume of phenol:chloroform:isoamyl alcohol (25:24:1; this can be purchased as a preprepared mixture) and vortex the sample to mix

Centrifuge the tube at 2 k rpm for 5 min (repeat the centrifugation if two distinct layers have not yet formed)

Remove the aqueous (top) layer and transfer it to a fresh microcentrifuge tube

Add one volume of phenol:chloroform:isoamyl alcohol (25:24:1) to the sample

Centrifuge for 5 min at 2 k rpm and remove the aqueous layer into a fresh microcentrifuge tube

Add 0.1 volume of sodium acetate (3 M pH 5.5)

Add 1 mL 100% ethanol or isopropanol (ice cold; you can at this point leave the sample at −20°C overnight) to precipitate the DNA

Centrifuge the samples at 13 k rpm for 30 min and pour away the ethanol/isopropanol—blot the tube gently on some paper towels to ensure that any residual solvent is removed

Wash the DNA pellet (you might not be able to see it, but you will have to trust it is there!) by adding 750 μL 70% ethanol to the tube and centrifuging for 1 min at 13 k rpm

Pour away the ethanol and blot as before

Air dry the pellet by leaving the lid of the tube open, at room temperature for 5—20 min

Resuspend the pellet in 50—100 μL Tris—EDTA (TE) buffer

Recipe for TE buffer (Prepare Using PCR Grade, Nuclease-Free Water)

10 mM Tris—HCl (pH, 7.5)

25 mM EDTA (pH, 8.0)

Once you have successfully extracted the template DNA you are ready to analyze and quantify it (see Chapter 2).

1.6 WHAT REAGENTS YOU WILL NEED AND WHERE TO BUY THEM

There are several essential reagents you will need to buy before you can run a PCR reaction, and there are many suppliers available who can supply the necessary reagents. The list below outlines the basics and serves as a tick-list to ensure you have everything you need.

Reagent Checklist

For PCR:

> **DNA extraction kit** (if using)
> **Glass beads** (if not using a kit)
> **Lysis buffer** (if not using a kit)
> **RNAse A, proteinase K** (if not using a kit)
> **Phenol:chloroform:isoamyl alcohol** (25:24:1; pH, 7.5−8 for DNA)
> **Ethanol** (100% and 70%)
> **Sodium acetate** (3 M; pH, 5.5—can be purchased or made)
> **Polymerase enzymes and associated buffers** (usually supplied as enzyme + tube of concentrated reaction buffer + Mg^{2+} solution)
> **Primers** (purchased as lyophilized powder you will need to reconstitute)
> **dNTPs** (mixed or individually)
> **Nuclease-free water** (or use nanopure water)
> **Reaction tubes** (normally nuclease-free PCR tubes)
> **Filter tips**

All of these reagents can be purchased from numerous suppliers; it is usually the best to see if anyone else has been working with PCR within your department, as there may be a preferred supplier. It is also always worth speaking to a few companies, to see which have the best deals on the products you are interested in. Some (but by no means all) of the companies you are likely to come across when shopping for these reagents are given below. Each will have their own regional website depending on the country you are in; you can link through to these from the webpages in the following list:

> **Bio-Rad**—www.biorad.com
> **Sigma**—www.sigmaaldritch.com
> **Qiagen**—www.qiagen.com
> **Thermo Scientific**—www.thermoscientificbio.com
> **Promega**—www.promega.com
> **GE Healthcare Lifesciences**—http://www.gelifesciences.com
> **Life Technologies**—http://www.lifetechnologies.com
> **Bioline**—www.bioline.com

1.7 SAFETY CONSIDERATIONS

It is of the utmost importance to consider the risks being taken when conducting any experiment, and it is no different with PCR. Basic safety protocols for your laboratory should be followed but in addition to this the material safety data sheet for every reagent you are using should be studied and you should be aware of how to safely store, use, and dispose of each reagent as well as the appropriate action to take should accidental spillage occur. It is always the best to label your reagents clearly with their identity if you have made aliquots into new containers, and the correct hazard labels should be attached to them in case anyone else has to deal with a spill. As such, it is imperative to handle all reagents wearing a lab coat, safety glasses, and gloves. You should also prepare a risk assessment containing each reagent you are using and display at your workstation for easy access.

SUMMARY

With the exception of your primer pairs, you will by now have designed, prepared, and ordered all of the things you need to undertake your first PCR and will also have your template DNA ready to use—this can be stored at either 4°C (short term) or −20°C (long term). In the next chapter, you will learn how to design primer pairs and analyze (quantify and assess purity) the template DNA and PCR product.

Chapter 2

Designing and Ordering Your Polymerase Chain Reaction Primers

ABSTRACT

This chapter provides practical advice on designing polymerase chain reaction primers and optimizing them. From using databases to identify genes and ordering primers, the reader is guided through the practical aspects of primer design and the common pitfalls. Guidance on how to test a new pair of primers is given along with a trouble-shooting guide.

2.1 INTRODUCTION

In order to amplify a gene of interest you need to know the gene sequence so that you can design primers. Primers are short sequences of DNA that are designed to be complementary to regions of your gene of interest; the area of DNA sequence in between the primers is the part that is amplified by polymerase chain reaction (PCR) and is referred to as the amplicon or PCR product. If your primers are not correctly designed, you can run into problems such as mispriming (where the primer binds to the wrong bit of DNA), multiple priming (where the primer binds many regions of DNA) or no priming at all (if the primers, for example, form a secondary structure rather than annealing to your template DNA).

Once you have designed and ordered your primers you must handle and store them correctly as they can degrade fairly quickly. Short pieces of DNA are ideal targets for DNases which can be inadvertently introduced into your primer stock. So always make sure you handle your primers in a "clean" way (as described in Chapter 1), wearing gloves, and using filter tips. Prepare a working stock of the primers (see Section 2.10) and store the master stock, relatively untouched.

This chapter will guide you through the basic parameters for primer design, handling, and preparation; by the end of it you should be able to confidently design PCR primers pairs that work!

2.2 CHOOSING A "TEMPLATE"

Before you can begin to design your primers you need to identify the target
gene or piece of DNA that you wish to amplify. Since the sequence of your
primers is fundamental to their success, where possible use gene sequence
from organisms that have had their genome fully sequenced, if not, it is
possible to identify DNA sequence for primer design using partially complete
genome sequence or contigs, which you can find on online databases (see
Section 2.3).

Decide the overall aim of your PCR; you might want to amplify just a short
sequence (try not to choose a target that is less than 100 bp), an entire gene, an
intergenic sequence, or several operonic genes (for prokaryotes). Very long
stretches of DNA (of 10 kb or above) can be difficult to amplify, but enzymes
exist that are specific for long PCR amplicons (see Chapter 1). With this in
mind, you must decide where your amplicon will begin and end, so essentially
the size of your amplicon. As a general rule of thumb, if you are trying to
amplify a whole gene it is a good idea to begin at the start codon and end at the
stop codon (Fig. 2.1), you might need to adjust the primer sites by a few base
pairs to account for annealing temperatures (see Section 2.7), but it is at least a

```
1       atgtctgcgg catccctgta ccccgtgcac ccggaagcgg tggcacggac cttcaccgac
61      gagcaggccc agcggatcga ctggatcaag ccgttcgaga aggtcaagca gacctccttc
181     gacgaccatc acgtggacat caagtggttc gccgacggca ccctcaacgt gtcgcacaac
241     tgcctcgacc gtcacctcgc cgaacgcggc gaccaggtgg cgatcatctg ggaaggcgac
301     gatcccgccg accaccagga aatcacctac cgccagctcc acgagcaggt ctgcaagttc
361     gccaacgccc tgcgcgggca ggacgtgcac cgcggcgacg tagtgaccat ctacatgccg
421     atgatccccg aggccgtggt cgccatgctc gcctgcaccc gcatcggcgc gatccactcg
481     gtggtcttcg gcggcttctc ccccgaggcc ctggcaggac gcatcatcga ttgcaagtcg
541     aaggtggtga tcaccgccga cgaaggcgtg cgcggcggca agcgcactcc gctgaaggcc
601     aatgtcgacg acgccctgac caacccggaa acctccacg tgcagaagat catcgtctgc
661     aagcgtaccg gtgcggagat caagtggaac cagcaccgcg acgtctggta cgacgacctg
721     atgaaggttg ccggcagcac ttgcgcaccc aaggaaatgg cgcgccgagga cccgctgttc
781     atcctctaca cctccggctc caccggcaag ccgaagggcg tgctgcatac caccggcggc
841     tacctggtgt acgcctcgct gacccacgag cgggtcttcg actaccgtcc gggcgaagtc
901     tactggtgca ccgccgacat cggctgggtc accggccaca cctacatcgt ctatggcccg
961     ttggccaacg gcgccaccac cattctgttc gagggcgtgc cgaactaccc cgacgtgacc
1021    cgcgtggcga aaatcatcga caagcacaag gtcaacatcc tctacaccgc gccgaccgcg
1081    atccgcgcga tgatggccga aggcaaggcg gcggtggccg gtgccgacg ttccagcctg
1141    cgtctgctcg gttcggtggg cgagccgatc aaccggaag cctggcagtg gtactacgag
1201    accgtcggcc agtcgcgctg cccgatcgtc gacacctggt ggcagaccga gaccggcgcc
1261    tgcctgatga ccccgctgcc gggcgcccac gcgatgaagc cgggctctgc agccaagccg
1321    ttcttcggcg tggtaccggc gctggtggac aacctcggca acctgatcga gggcgcgcc
1381    gagggcaacc tggtgatcct cgactcctgg ccgggccagg cgcggaccct gttcggcgac
1441    catgaccgct tcgtcgacac ctacttcaag accttcaagg gcatgtactt caccggcgac
1501    ggcgcgcgcc gcgacgagga cggctactac tggatcaccg ggcgggtcga cgacgtgctc
1561    aacgtctccg gccaccgcat gggcaccgcc gaggtggaaa gcgcgatggt cgcccacccg
1621    aaggtcgccg aggcggcggt ggtcggcatg cagcacgaca tcaagggcgca gggcatctat
1681    gtctacgtca ccctcaactc cggggtcgag ccgagcgagg cgctgcgcca ggagctgaag
1741    caatgggttc gccgcgagat cggtccgatc gccacgcgcgg atgtgatcca gtgggcgccg
1801    ggactgccga agacccgttc gggcaagatc atgcggcgca tcctgcgcaa gatcgccgcg
1861    gccgagtacg acaccctcgg cgacatctcc acccttgccg acccgggcgt ggtccagcac
1921    ctgatcgaga cccatcgctc gatgcaggcc gcctga
```

FIGURE 2.1   Gene sequence showing primers designed to prime at the beginning and end of the
gene, thus amplifying the entire gene.

start. On this figure, the start and stop codon are highlighted in bold, and the selected primer sequence is highlighted in yellow. If you are not amplifying an entire gene, decide the length of your required amplicon and pick an area that you would like to be your forward primer and another that will be your reverse primer—you will invariably make adjustment to this initial choice, but again, this gives you something to start with.

The G+C content of the genome you are working with can cause problems for primer design. An "ideal" organism will have a typical G+C ratio of about 50%. Organisms with high G+C ratios mean that you can end up with primers that have very high annealing temperatures. Try to avoid regions of DNA that are highly repetitive or have runs of any one base—this can be tricky if you are attempting to amplify an intergenic region of sequence. If the odds are stacked against you in terms of the sequence you have to work with, then design the best primer pair that you can and then adjust your PCR parameters to maximize your chances of success (see Section 2.12 for common problems and troubleshooting).

2.3 FINDING GENE SEQUENCES USING ONLINE DATABASES

This is all very well, but how exactly do you find the sequence you want to work with? There are a number of online databases you can use. If you are lucky, there might be a designated genome database for the organism you are interested in, for example, for *Pseudomonas aeruginosa*, the database www.pseudomonas.com is very useful. Alternatively, the NCBI database is a good place to look (www.ncbi.nlm.nih.gov). This section of the chapter will assume that there are no specific databases and will guide you through how to use the NCBI database to find DNA or gene sequence.

On the NCBI homepage (Fig. 2.2), select "DNA & RNA" from the blue menu on the left-hand side of the screen. From the list of databases that appear, select "Genbank". Genbank is the National Institute of Health genetic sequence database, which provides an annotated collection of all DNA sequences that are currently publicly available. You have the option of searching Genbank using "Entrez Nucleotide" or basic local alignment search tool (BLAST), depending on the information that you have regarding the sequence you wish to amplify.

If you have a sequence identifier, i.e., an annotated gene name or locus number, you can type it into the search box in nucleotide. If you have a nucleotide sequence but not annotation, you can use BLAST. For nucleotide, simply type the gene name or locus number into the search box (Fig. 2.3) and click search. It is often helpful to put additional information in here to narrow the results, such as the name of the organisms and the subspecies or strain, e.g., *Escherichia coli* K12 *fur* (the bacterial name, its strain designation, and the name of the gene of interest).

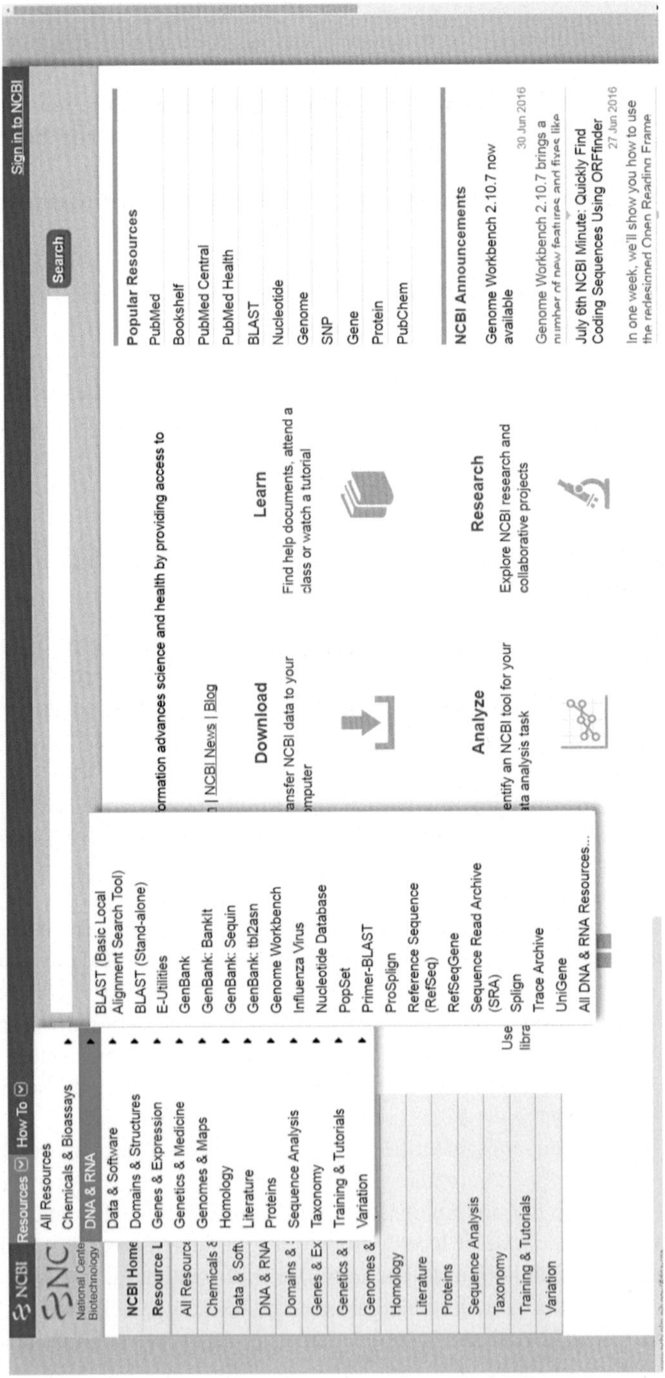

FIGURE 2.2 NCBI homepage and where to find Genbank.

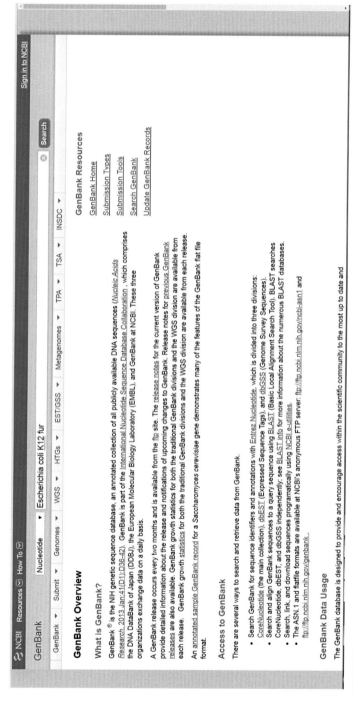

FIGURE 2.3 Searching a nucleotide sequence. Note the name of the organism and gene of interest has been entered into the search box at the top of the page.

Your search terms will most likely bring up a number of completed genome sequences, for example, by searching the terms above, the hits returned included a number of *E. coli* that had been fully sequenced but also strains and bacteria other than K12 (Fig. 2.4).

Scroll down the list until you find the correct information, and click on the link. This opens a page with the basic genome information on it, such as the source and associated publications (Fig. 2.5A)—scroll down and you will find the annotated genome as a list of gene names or locus designations, their start and end sites within the genome (numbered), and the gene function.

You can simply search for your gene of interest using the "find" function on any web browser. You will notice links next to the gene information ("gene" and "CDS" (coding DNA sequence); Fig. 2.5B)—if you click on the "gene" link a new page will open containing the gene sequence. So now you can use this to begin designing your primers! If you want to design primers for intergenic regions, then note down the base-pair locations of your genes, then scroll down to the very bottom of the page where the entire genome is given as one long run of nucleotides. Each line is numbered so you can find the location of your genes and in doing so also find the intergenic regions.

If you have a sequence of DNA that is not annotated, you can use BLAST to help to identify it and to allow you to target a specific region of that gene. You can also use BLAST to see how many other organisms contain your gene of interest, and how conserved the gene sequence is. By doing this you can work out whether your primer pair could, for example, be used to amplify the same target from a number of different organisms, or at least if you could design them to anneal to a conserved region of an otherwise variable gene within a group of organisms.

To use BLAST, click on the BLAST link (Fig. 2.6).

On the BLAST webpage you will notice, there is a list of groups or families or kingdoms of organisms you can choose from, for e.g., human, mouse, rat, and cow. If your sequence falls within any of these subgroups, then use these links to aid with your search. If not, select nucleotide BLAST. Simply paste your sequence (in FASTA (FAST-All) format, i.e., one continuous, non-annotated, unnumbered, unbroken sequence) into the box and hit submit. If your organism of interest appears in the drop down menu of the box entitled "organism," then select this to help to refine the search. Your BLAST results will be presented as a series of "hits" with decreasing percentage identity (Fig. 2.7). If you are lucky, you will hit on an annotated gene.

A newer function of BLAST is "primer-BLAST" (Fig. 2.8). By pasting your nucleotide sequence into the box and specifying the parameters for your primer, you can get the program to design the primers for you. The program will also predict amplicon size and annealing temperature as well as primer pair specificity.

The approach you opt for depends on what you are trying to amplify and everyone generally has their own preferred method of finding DNA sequence

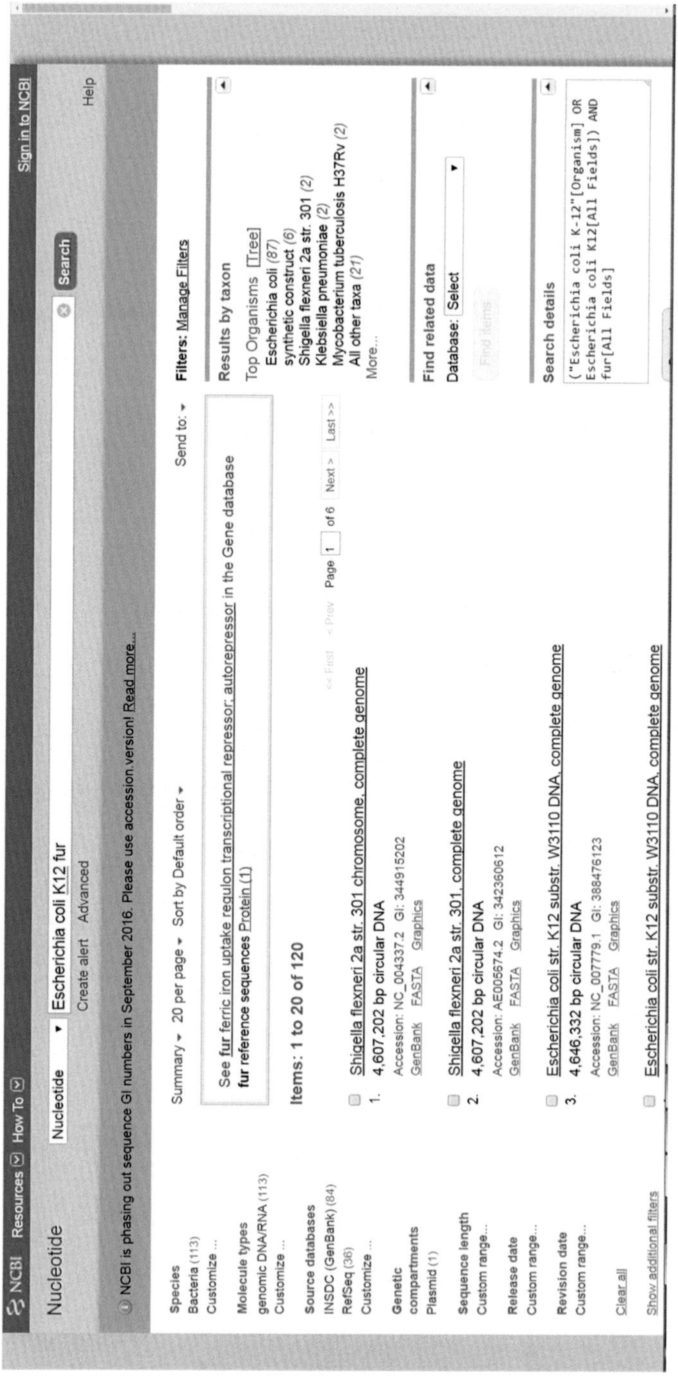

FIGURE 2.4 Nucleotide search results for *Escherichia coli* K12 *fur.*

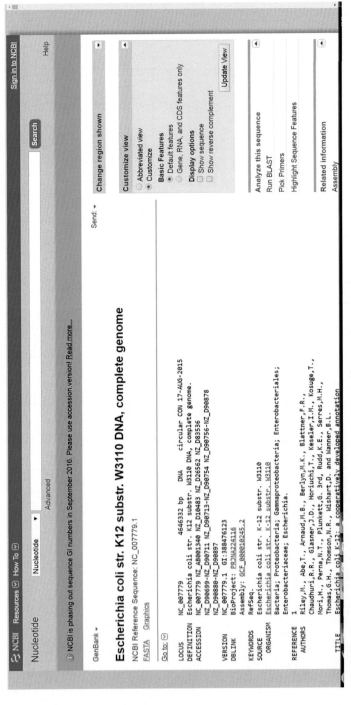

FIGURE 2.5A Basic genome information.

FIGURE 2.5B Searching through gene information.

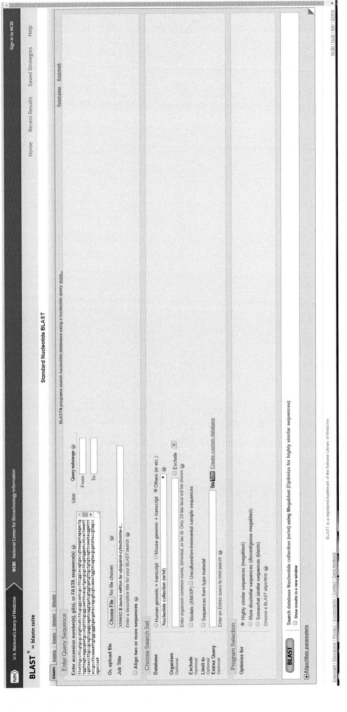

FIGURE 2.6   The BLAST webpage. Note the sequence pasted into the query box and the different parameters (beneath) that can be changed accordingly.

FIGURE 2.7 BLAST hits.

Descriptions

Sequences producing significant alignments:

Select: All None Selected:0

Alignments Download GenBank Graphics Distance tree of results

Description	Max score	Total score	Query cover	E value	Ident	Accession
Escherichia coli isolate E. coli RL465 genome assembly, chromosome: RL465_chromosome	826	826	100%	0.0	100%	LT594504.1
Escherichia coli strain ER1821R, complete genome	826	826	100%	0.0	100%	CP016018.1
Escherichia coli str. Sanji, complete genome	826	826	100%	0.0	100%	CP011061.1
Escherichia coli str. K-12 substr. MG1655 strain JW5437-1, complete genome	826	826	100%	0.0	100%	CP014348.1
Escherichia coli K-12 strain C3026, complete genome	826	826	100%	0.0	100%	CP014272.1
Escherichia coli K-12 strain DHB4, complete genome	826	826	100%	0.0	100%	CP014270.1
Escherichia coli str. K-12 substr. MG1655, complete genome	826	826	100%	0.0	100%	CP014225.1
Escherichia coli strain SEC470 genome	826	826	100%	0.0	100%	CP013962.1
Escherichia coli strain ACN002, complete genome	826	826	100%	0.0	100%	CP007491.1
Escherichia coli strain CQSW20, complete genome	826	826	100%	0.0	100%	CP013253.1
Escherichia coli strain YD786, complete genome	826	826	100%	0.0	100%	CP013112.1
Escherichia coli strain 2012C-4227, complete genome	826	826	100%	0.0	100%	CP013029.1
Escherichia coli K-12 GM4792 Lac-, complete genome	826	826	100%	0.0	100%	CP011343.2
Escherichia coli K-12 GM4792 Lac+, complete genome	826	826	100%	0.0	100%	CP011342.2
Escherichia coli strain K-12 substrain MG1655 TMP32XR2, complete genome	826	826	100%	0.0	100%	CP012870.1
Escherichia coli strain K-12 substrain MG1655 TMP32XR1, complete genome	826	826	100%	0.0	100%	CP012869.1
Escherichia coli str. K-12 substr. MG1655, complete genome	826	826	100%	0.0	100%	CP012868.1
Escherichia coli genome assembly ERS742059, chromosome :I	826	826	100%	0.0	100%	LN877770.1
Escherichia coli strain RR1, complete genome	826	826	100%	0.0	100%	CP011113.1
Escherichia coli strain DH1Ec169, complete genome	826	826	100%	0.0	100%	CP012127.1
Escherichia coli strain DH1Ec104, complete genome	826	826	100%	0.0	100%	CP012126.1
Escherichia coli strain DH1Ec095, complete genome	826	826	100%	0.0	100%	CP012125.1
Escherichia coli ACN001, complete genome	826	826	100%	0.0	100%	CP007442.1
Escherichia coli strain NCM3722, complete genome	826	826	100%	0.0	100%	CP011495.1
Escherichia coli strain 94-3024, complete genome	826	826	100%	0.0	100%	CP009106.2
Shigella boydii strain ATCC 9210, complete genome	826	826	100%	0.0	100%	CP011511.1
Escherichia coli strain SQ2203, complete genome	826	826	100%	0.0	100%	CP011324.1
Escherichia coli strain SQ171, complete genome	826	826	100%	0.0	100%	CP011323.1

FIGURE 2.7 BLAST hits.

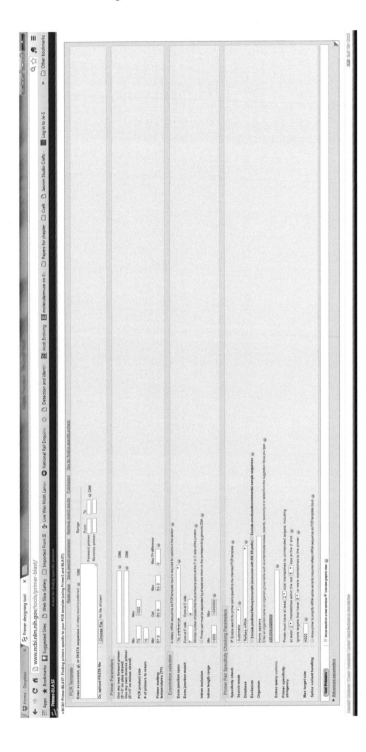

FIGURE 2.8 Primer BLAST.

and subsequent primer design. The descriptions above are simply a few good start points.

2.4 IDENTIFYING YOUR AMPLICON OF INTEREST

So now, you should have some sequence to begin work with. If you want to amplify the entire gene, then one of your primers will be designed to anneal at the start and the other at the end (Figs. 2.1 and 2.2). If this is not what you aim to achieve, then when you acquire your gene sequence from the database, get a bit of extra sequence upstream and downstream of the gene to allow for "design room"; if you are amplifying an internal segment of the gene you would not need this. By doing a bit of sequence gazing, you can look to see if there are any runs of identical bases or any regions of the gene with high numbers of G+C, these are regions to avoid when designing your primers.

If you are working with a primer-design program (see Section 2.8), then copy and paste the sequence directly into the program (use FASTA format); if not you can copy and paste it into a text file for use later, so that you do not have to find it all over again.

2.5 CONTENDING WITH INTRONS AND EXONS

If you are working with prokaryotic organisms, then this section is not for you, you can skip straight onto Section 2.6. If you are trying to PCR amplify genes from a eukaryotic organism, you must bear in mind that there can be differences between the predicted sizes of you gene product using the sequence, and what you get following PCR. Introns cause this discrepancy. If you are amplifying directly from a piece of genomic DNA, then the amplicon size and the size of the nucleotide sequence you are using to design your primer will be the same. If you are using cDNA (Chapters 6 and 7) as a template, the amplicon size will be smaller than the genomic DNA sequence as introns are spliced out. When using cDNA as a template, it is a good idea to identify intron-exon boundaries and design at least one of your primers to span the spliced region, so that they will anneal only to cDNA and not genomic DNA. If it's not possible to do this you can always treat your sample with a dsDNase prior to amplification.

2.6 BASIC PRIMER PARAMETERS

A successful PCR reaction yielding the optimal amount of product requires specific and efficient amplification of your target. The target you choose and the primers you design can both affect the efficacy of your reaction. So, there are a few basic rules to bear in mind when embarking on primer design that will maximize your chances of successfully creating functioning primer sets.

Hint: Before you start designing your own primers, check the literature. It may well be that someone has already designed primers that will do the job for you!

If you do design your own primer sets, then if you follow these basic parameters then hopefully when you first test out your new primers they will produce the expected products.

- Primer length: The length of your primer is going to affect both its specificity and the annealing temperature. Optimal size is between 18 and 25 nucleotides to avoid issues with specificity and efficiency. The purpose of PCR primers is to provide a "free" 3′-OH group to which the DNA polymerase can add dNTPs.
- Melting temperature (Tm): There are two primers in your reaction for both to have maximum efficiency, and it is important that their melting temperature be as similar as possible (within 2°C if possible). There are various online programs that can help with design of primers and calculating Tm, (some of the design sites will not give the same values as the site you order your primers from; use the Tm as a guide to determine the annealing temperature rather than an absolute value). If you are not using a program to design your primers, you can estimate the melting temperature as indicated in the equation:

$$Tm \ (^{\circ}C) = 2 \ (A+T) + 4 \ (G+C)$$

Using this you should aim to design primer with a Tm of between 60°C and 64°C.

- GC (guanine and cytosine) content: Your primers should have a GC content of between 50% and 55%. When calculating this, it is also important to bear in mind that you do not want long repeats of any nucleotide. Generally, greater than three bases of one type, for e.g., TTT should be avoided where possible. This will help minimize the chance of secondary structures occurring.
- Complementary primer sequence: The primer you design should not have more than three base pairs of intraprimer homology to avoid getting double-stranded structure (DS) formation (DS structure presence will interfere with the efficacy of annealing). The same is true of the primer pair; if possible they should not contain homologous regions which may cause primer dimer or hairpin formations.

At its most basic, your forward primer is a short segment of DNA sequence that is identical to the "top-strand" of DNA (Fig. 2.9)—DNA sequence is given as one strand only, the "top-strand" and complementary bases are not given. If you are using primer-design software and have pasted the sequence into it, there is usually an option to enable you to see the complementary sequence. If you are not able to do this, choose your reverse primer by selecting a short segment from the sequence you have, then note down the complementary bases, and finally turn it around so that it runs in the opposite direction (reverse complemented; Fig. 2.9).

Forward   GCGACGGTATTCGAACTTGT
Reverse   CGAAGAAGGAAATGGTCGAG
Tm
Amplicon length

Sequence
```
634321 ggcaagcggc aaagaaaaac ccccggtttt tcaggccggg ggttcgttcg tcactcgtcg
634381 aggaaggagc gaaggtgctc gcttcgcgac ggatggcgca gcttgcgcag cgccttggct
634441 tcgatctggc ggatccgctc gcgggtcaca tcgaactgct tgccgacttc ctcgagggtg
634501 tggtcggtgt tcatgtcgat gccgaagcgc atgcgcagca ccttggcttc ccgggcagtg
634561 aggccggcga ggacttcgcg ggtggattcc ttgaggctct cgctggtcgc catctcgatc
634621 ggcgactgca tggtggagtc ctcgatgaaa tcgcccaggt gcgaatcttc gtcgtcaccg
634681 atcggggttt ccatggagat cggctctttg gcgatcttca gtaccttgcg gatcttgtcc
634741 tcaggcatgt ccatgcgctc gccaagctct tccggggtgg gctcgcgacc catttcctgg
634801 agcatctggc gggagatgcg gttgagcttg ttgatcgtct cgatcatgtg caccgggatg
634861 cggatggtgc gtgcctggtc ggcgatcgaa cgggtgatcg cctggcgaat ccaccaggtg
634921 gcgtaggtgg agaacttgta gccgcgacgg tattcgaact tgtccaccgc cttcatcagg
634981 ccgatgttgc cttcctggat caggtcgagg aattgcaggc cgcggttggt gtacttcttg
635041 gcgatggaaa tcaccaggcg caggttggcc tcgaccattt ccttcttcgc ccggcgagcc
635101 ttggcttcgc cgatcgacat cgcgcggttg atttccttga tctcggcgac ggtcagctcg
635161 acctcgcttt ccagggccgc cagcttctgc tggttgcgca ggatgtcgtc gcgcaggcgc
635221 tcgatggcct cggcgtactt cggcttgctc ttcaggacgc tgtcgaccca cttctcgtcg
635281 gtctcgtggt tcgggaacag gcgcaggaag tcggcacgcg gcatgcgcgc gtcacgcacg
635341 cagagctgca tgatggcgcg ttcctgggcg cgcacgcctt ccagggcgga gcgcacgcgg
```

FIGURE 2.9 Forward and reverse DNA primers with priming sites highlighted on the gene sequence in green.

Having chosen the sequence from which to design your primer pair (see Section 2.4) and having selected regions for the primer sequence, it is a good idea to note exactly where they will bind on your region of DNA of interest, and then work out the expected size of the amplicon (the sequence in between and including your primers). Make a note of this size, as it will be useful when you electrophorese you PCR product and is useful to keep in your lab book (Fig. 2.10).

Forward   GCGACGGTATTCGAACTTGT
Reverse   CGAAGAAGGAAATGGTCGAG
Tm
Amplicon length   147 bp

Sequence
```
634321 ggcaagcggc aaagaaaaac ccccggtttt tcaggccggg ggttcgttcg tcactcgtcg
634381 aggaaggagc gaaggtgctc gcttcgcgac ggatggcgca gcttgcgcag cgccttggct
634441 tcgatctggc ggatccgctc gcgggtcaca tcgaactgct tgccgacttc ctcgagggtg
634501 tggtcggtgt tcatgtcgat gccgaagcgc atgcgcagca ccttggcttc ccgggcagtg
634561 aggccggcga ggacttcgcg ggtggattcc ttgaggctct cgctggtcgc catctcgatc
634621 ggcgactgca tggtggagtc ctcgatgaaa tcgcccaggt gcgaatcttc gtcgtcaccg
634681 atcggggttt ccatggagat cggctctttg gcgatcttca gtaccttgcg gatcttgtcc
634741 tcaggcatgt ccatgcgctc gccaagctct tccggggtgg gctcgcgacc catttcctgg
634801 agcatctggc gggagatgcg gttgagcttg ttgatcgtct cgatcatgtg caccgggatg
634861 cggatggtgc gtgcctggtc ggcgatcgaa cgggtgatcg cctggcgaat ccaccaggtg
634921 gcgtaggtgg agaacttgta gccgcgacgg tattcgaact tgtccaccgc cttcatcagg
634981 ccgatgttgc cttcctggat caggtcgagg aattgcaggc cgcggttggt gtacttcttg
635041 gcgatggaaa tcaccaggcg caggttggcc tcgaccattt ccttcttcgc ccggcgagcc
635101 ttggcttcgc cgatcgacat cgcgcggttg atttccttga tctcggcgac ggtcagctcg
635161 acctcgcttt ccagggccgc cagcttctgc tggttgcgca ggatgtcgtc gcgcaggcgc
635221 tcgatggcct cggcgtactt cggcttgctc ttcaggacgc tgtcgaccca cttctcgtcg
635281 gtctcgtggt tcgggaacag gcgcaggaag tcggcacgcg gcatgcgcgc gtcacgcacg
```

FIGURE 2.10 Noting amplicon size.

2.7 CALCULATING ANNEALING AND MELTING TEMPERATURES

The melting temperature of flanking primers should not differ by more than 2−5°C to ensure maximum efficiency of your reaction. Therefore, the GC content and length must be chosen carefully. A general rule of thumb for estimating the melting temperature has been given in Section 2.6; if you use this formula be aware that it is likely to give you a slightly different value to primer-design programs, and indeed to the values provided with the documentation when you receive your primers. This is because a number of different algorithms can be used, and therefore, these values are used primarily as a guide to help to estimate the appropriate annealing temperature. Such algorithms take into consideration other factors such as interactions between adjacent bases and salt concentration.

When running your PCR reaction, the annealing step commonly lasts for 30−45 s, giving enough time for primers to locate their complement and anneal to the template DNA. Ideally, the annealing temperature (Ta) should be approximately 5°C lower than your melting temperature. However, if nonspecific PCR products are obtained in addition to the expected product, the annealing temperature should be optimized by increasing it stepwise by 1−2°C both above and below your calculated Ta (it is often a good idea to use a temperature gradient; Section 2.11).

2.8 TOOLS FOR PRIMER DESIGN

There are many tools for helping with primer design. These can be purchased as software packages that require licensing for use, or else there are independent websites where you can design primers (Fig. 2.11) and assess their specificity.

If you already have you sequence or want to design primers using programs provided by your chosen supplier, you simply have to enter the information into the online form. You will normally be asked to define the parameters covered above, giving enough information so that primer pair can be derived; often this uses a set up similar to the one shown in the table.

Primer Size			Primer Tm °C			Primer GC%		
Min	Optimal	Max	Min	Optimal	Max	Min	Optimal	Max
18	20	27	57	60	63	20	50	80

Region of Analysis		Product Size (bp)		Experimental Conditions	
From	To	Min	Max	Salt Conc	Primer Conc
1	=The length of your sequence	100	300	50 mM	50 nM

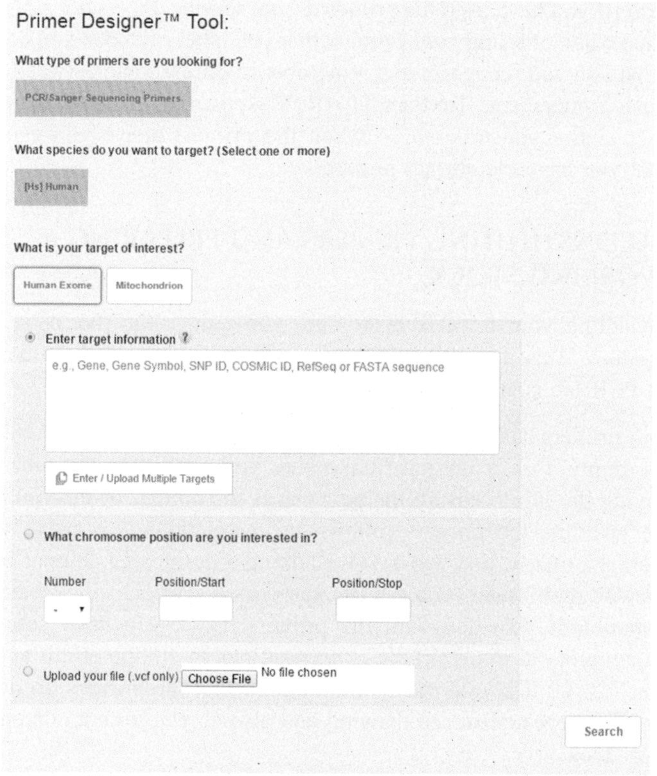

Primer Designer™ Tool:

What type of primers are you looking for?

PCR/Sanger Sequencing Primers

What species do you want to target? (Select one or more)

[Hs] Human

What is your target of interest?

Human Exome | Mitochondrion

○ Enter target information ?

e.g., Gene, Gene Symbol, SNP ID, COSMIC ID, RefSeq or FASTA sequence

Enter / Upload Multiple Targets

○ What chromosome position are you interested in?

Number   Position/Start   Position/Stop

- ▼

○ Upload your file (.vcf only) Choose File   No file chosen

Search

FIGURE 2.11   Online primer designing software.

Once you have provided this information you can select how many primer pairs you want—you can then choose your preferred primer pair form those suggested. Most sites will automatically generate several sets of primer pairs for you to choose from.

2.9 ORDERING YOUR PRIMERS

There are numerous on line suppliers of primers, many of which also offer primer designing services. Some of the most well-known include:

Sigma-Aldritch—http://www.sigmaaldrich.com
Life Technologies—http://www.lifetechnologies.com
Biorad—http://www.bio-rad.com
Eurofins—http://www.operon.com

Ordering your primers should be a fairly straightforward process and mostly relies on you making sure that you enter the correct sequence into the order form (if you have used the primer-design service, this is not a problem!). Ensure that when ordering your primers that you label each one you order with a name that you will recognize and remember to differentiate between forward and reverse primers, e.g., ToxF and ToxR. Make sure you keep a record of the primer sequences you ordered so when the primers arrive you can double check that you have the correct primer.

2.10 RECONSTITUTING PRIMERS AND PREPARING A WORKING STOCK

Before handling your primers make sure you reduce the risk of accidently contamination. Wear a clean laboratory coat (you can have dedicated pre-PCR and post-PCR lab coats) and always put on clean gloves.

- If your primers have come in the lyophilized form (do not panic if the vial looks empty, that is normal) make sure you centrifuge the tubes before removing the lid, to ensure the pellet is at the bottom of the vial.
- To reconstitute the primers follow, the instructions that arrive with the primers. Normally, this will involve adding a designated amount of sterile molecular grade water (you can use autoclaved sterile distilled water). This information is provided with the primers. Once you have your master stock, aliquot out into working concentrations of 100 pmol/mL and freeze in the −20°C. It is important to aliquot so that the primers do not break down after repeated freeze thawing and also so you do not contaminate a master stock.

2.11 TESTING A NEW PRIMER PAIR

Once you have reconstituted and aliquoted your primers, it is important to make sure they are working and specific for your target gene—for most people this will be the first PCR, so now is a good time to check back over your list of reagents and "PCR-ingredients" as prepared in Chapter 1.

If you have access to a PCR machine with a temperature gradient, it is recommended to use this for the first reaction. Calculate the expected Ta and then pick a range of temperatures that encompass 5°C higher and 5°C lower than the Ta. For example, if the Ta is 55°C, then the gradient should run between 50°C and 60°C. Most machines with a temperature gradient facility will automatically calculate the individual reaction temperatures, and you just have to provide the highest and lowest temperature you wish to use (Fig. 2.12). Typically, primers are added to a reaction at concentrations anywhere between 0.2 and 0.55 μM. However, you can always try a range of primer concentrations from 0.1 to 1 μM.

Having decided upon the Ta (or selected to use a gradient) you need to program the remaining cycle parameters (see the manufacturers instructions

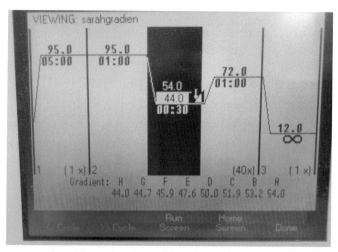

FIGURE 2.12   Typical PCR reaction parameters entered into a PCR machine. Note the extension step has a gradient with temperature for different wells given along the bottom of the screen.

depending on what machine you are using). A standard thermal cycling profile will look something like this:

Initial denaturation step—95°C for 5 min (done just once at the beginning of the reaction)
Denature—95°C for 30—60 s
Annealing—X°C for 30—45 s
Extension—70°C for 1—2 min
Repeat steps two to four for between 25 and 40 cycles then add a final extension step of 70°C for 8—10 min.

If you are not around when your PCR finishes, it is important to program the machine to hold the PCR products at 4—12°C until you can remove them from the machine (for example, you might set the reaction to run overnight). It is also a good idea to stick a note to both the machine and the wall plug to let people know the machine is running and not to turn the power off. There is nothing more frustrating than finding out your PCR failed because someone accidently turned it off half-way through!

General Notes

If possible, try to use a PCR machine that has a heated lid to stop your sample mixture form condensing into the lid of your reaction tube; if you do not have access to one then overlay each reaction with liquid paraffin.

It might sound obvious, but it always a good idea to have both positive and negative controls. You should always a have a DNA template free reaction (negative control) to ensure there is no contamination within your reaction mixture. You should get a clear lane with no banding when you run this sample on your gel. A positive control can be purchased from any company that provides molecular reagents (as previously listed) or can be borrowed from a lab mate!

2.12 COMMON PROBLEMS AND TROUBLESHOOTING

Even the best designed PCR reactions can run into problems and below are a few you might encounter, with possible causes and solutions (see Chapter 3 for analysis and interpretation of your PCR reaction).

Problem	Possible Causes and Solutions
No product	1. Insufficient DNA added. (Try DNA extraction again, quantify amount if in doubt about yield, try adding range of DNA concentrations)
	2. Primer not added (ensure you spot small volumes onto inside of tube so you can see you have added them, mix contents thoroughly before adding to PCR machine.
	3. Mastermix concentrations wrong (check ratios of ingredients to ensure they are correct)
	4. Impure DNA (extract DNA again or heat sample to 95°C for 5 min before use in PCR machine
	5. Thermal cycler not functioning properly (test again with known primers to check)
Background amplification	1. Too much DNA added to reaction (reduce amount of DNA added until background contamination reduced)
	2. Too many cycles were used (reduce number of cycles used)
	3. Too much primer (reduce amount of primer used, can use titration)
	4. Improper annealing temperature (check annealing temperature of primers, set a gradient of temperatures in 1°C steps)
	5. Poor quality DNA (isolate new DNA)
	6. Poorly designed primers (check there is no secondary structure or complementarity between primers, redesign if necessary)
Nonspecific amplification	7. Primers hybridizing to more than one site on template (increase annealing temperature by gradient, or design new primers)
	8. Contamination! make new stocks of primers/buffers/DNA template

Chapter 3

Analyzing Your Template DNA and/or PCR Product

ABSTRACT

This chapter explains how to analyze your template DNA and end point polymerase chain reaction (PCR) reaction using classical electrophoresis techniques as well as by spectrophotometry. It explains how to interpret what you see to determine yield and purity of the DNA/PCR product. This chapter also offers useful recipes for basic reagents necessary to carry out analysis of PCR by electrophoresis, which can be adapted to the user's needs.

3.1 INTRODUCTION

To verify whether your chromosomal extraction and/or polymerase chain reaction (PCR) have been successful, the product can be analyzed using gel electrophoresis or spectrophotometry. Electrophoresis enables separation of DNA fragments according to their size through a gel matrix. This will give you an indication of whether the extraction and/or reaction has worked, and also how much template DNA/PCR product you have, as well as giving an indication of its purity. Spectrophotometry is usually reserved for quantification of extracted DNA template or RNA, rather than analysis of PCR products. The absorbance readings can be used to indicate the yield and purity of either DNA or RNA prior to downstream applications such as quantitative or reverse-transcriptase PCR (see Chapter 4). We will begin by considering electrophoresis, a highly versatile method with a number of variations in terms of equipment and setup.

3.2 BASIC DNA/PCR ANALYSIS

Having successfully extracted your template DNA or amplified your target of choice, it is now equally important to be able to analyze it. Gel matrices comprised of agarose dissolved in an electrophoretic buffer are utilized to achieve this. The process of gel electrophoresis relies on an electrical current to pull the negatively charged DNA through the agarose gel matrix, toward the positively charged cathode. Smaller molecules move more slowly through the sieve-like gel matrix and therefore migrate further through the gel than the larger fragments, producing a distinct series of "bands." Therefore, the concentration of the agarose gel can have quite an impact on how the DNA fragment resolve, and this must be considered before casting the gel! You can analyze electrophoresed DNA by densitometry to give an indication of how much is present.

3.3 "RUNNING" AN AGAROSE GEL

To analyze extracted DNA/PCR products by electrophoresis the first thing you will need to make is an agarose gel, these can be purchased precast from a number of suppliers but are also very easy (and relatively cheap!) to make. It is important that you have an idea of the size of DNA fragment you want to analyze as this informs the percentage composition of the gel you prepare. Table 3.1 is a guide showing the percentage (w/v) of agarose required to resolve different sized fragments of DNA. The higher the percentage of agarose you use in your gel, the smaller the spaces of the "agarose-sieve" become, therefore making it increasingly difficult for large DNA molecules to travel through the gel.

Before you begin, calculate the volume of gel your gel tray will hold—small gel trays typically hold 30 mL and medium ones 50 mL but test by pouring water into the tray. Next, calculate the amount of agarose you need to weigh out as a percentage of the total volume of buffer. Transfer the appropriate volume of buffer (Table 3.2) into a conical flask and add the agarose powder to it. There are a number of different buffers to choose from, and it does not really matter too much which you choose; however, TBE is a better conductive medium than TAE and so is better suited for longer runs of electrophoresis (and is also purported also to give better resolution or sharper banding). TAE is considered to be better if the DNA being analyzed will be

TABLE 3.1 Percentage of Agarose Gel to Prepare for Resolution of Different Sized DNA Fragments

% (w/v) Agarose	Optimum Resolution (kb)
0.5	1–30
0.7	0.8–12
1.0	0.5–10
1.2	0.4–7
1.5	0.2–3
2.0	0.05–2

TABLE 3.2 Buffers for Electrophoresis

Buffer	Working Solutions	Stock Solutions
TAE Tris-acetate-EDTA	1× 40 mM Tris acetate 1 mM EDTA	50× 242 g Tris base 57.1 mL glacial acetic acid 100 mL 0.5 M EDTA (pH 8)
TBE Tris-borate-EDTA	0.5× 45 mM Tris acetate 1 mM EDTA	5× 54 g Tris base 27.5 mL boric 20-mL 0.5 M EDTA (pH 8)
TPE Tris-phosphate-EDTA	1× 90 mM Tris acetate 2 mM EDTA	10× 108 g Tris base 15.5 mL Phosphoric acid (85%, 1.679 g/mL) 40 mL 0.5 M EDTA (pH 8)

used in downstream applications, such as cloning, as there is less carry-over of salts, which can impair processes such as ligation.

All of these buffers can be stored at room temperature. If you make these buffers up at higher concentrations than the stock solutions suggested above, you will find that the solutes start to precipitate during storage. It is best to check for precipitates around the tap/neck of the container you are keeping your stock in before using the buffer; if there is a significant amount of precipitation it will alter the concentration of dissolved salts.

The agarose-buffer mixture needs to be heated to dissolve the powdered agarose, and this can be done either over a Bunsen or in a microwave (low power for periods of 1 min, with swirling). In either case keep a **very** close eye on the mixture as if overheated, it will boil over and make sure you wear eye protection and heat-proof gloves. When heating agarose, invert a second, smaller, conical flask inside the top of the one containing the mixture, this will ensure that any steam rising off the gel will condense and run back in; if you do not do this you can lose too much liquid from the mixture which affects the concentration. While heating over a Bunsen burner you need to periodically swirl the container until it is near boiling, then remove from the heat, and look carefully to see if the mixture is completely clear with no specks of undissolved agarose remaining.

Once the mixture has been adequately heated and looks clear, place the flask in a water bath at 55°C, wait until it has cooled, and there are no bubbles, before pouring the gel. Pouring while there are bubbles can lead to uneven gel surface that will make it harder to analyze later (see Section 3.4 for staining DNA in situ).

Most gel tanks will provide a variety of combs with a different number of teeth (wells) so before you pour the gel, decide how many wells you will need, do not forget to leave room for a molecular weight marker and your controls (Fig. 3.1).

Setup the casting equipment with the chosen comb in it and smoothly pour the gel into it; you should see the gel going between and behind the combs teeth; you do not want to introduce air bubbles—your gel should be between 3 and 5 mm thick (Fig. 3.2).

Some electrophoresis systems come with "dams" to seal the ends of the casting tray, if yours does not, and then seal the ends well with masking tape.

Once the gel is set (roughly 30 min), but before you add the running buffer to the tank (the same buffer as you made your gel with) make sure you have orientated your gel in the correct direction so that the wells are at the anode (negative) end of the tank (remember DNA is negatively charged and so runs to the cathode—if the gel tank is color coded, remember that DNA "runs to red"). Pour in your running buffer until it is just covering the gel then gently remove the comb from the gel so the wells are ready to fill with the samples (Fig. 3.3).

Before loading the samples you will need to mix them with some loading dye, this increases the density of the samples helping them settle to the bottom

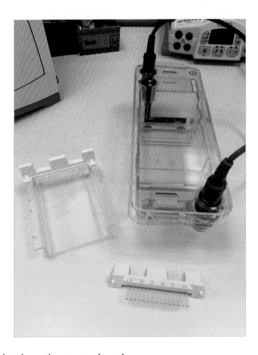

FIGURE 3.1 Gel tank, casting tray, and comb.

FIGURE 3.2   Gel preparation. The casting tray has been sealed with tape, and the agarose poured in.

FIGURE 3.3   Gel in place in the electrophoresis tank with wells at the anode end of the tank.

of the wells. The color of the dye will also make it easier to see when you are applying your samples to the wells. You will dilute the stock loading dye to the working concentration with your PCR product, so for example, 5 µL PCR product plus 1 µL of loading dye. Table 3.3 gives some common recipes for loading dyes.

All of these dyes need to be stored at 4°C. With the dyes, as with the buffer which one you chose is a matter of personal preference and often influenced by what people in the lab have used before. The dyes "run" at different positions on the gel, for example, bromophenol blue runs at approximately 1 kb in a 0.8% agarose gel, whereas xylene cyanol runs at approximately 10 kb. When visualized using UV light, these dyes can produce a "shadow" making DNA bands of a similar molecular weight difficult to see. Orange G does not produce this shadowing effect and runs at <100 bp.

One you have mixed your sample with the loading dye, pipette 5—10 µL into the wells of the gel. This can be tricky at first, and if you have trouble seeing where the wells are you can place a piece of black paper underneath the gel tank (Fig. 3.4).

Having applied your samples to the gel (including the molecular weight marker), you now need to separate them by passing an electrical current through the system. You should run your gel at 1—5 V/cm (measure the distance between the cathode and anode); when the electrical current is applied, you will see the dye beginning to move through the gel bubbles being produced at the cathode. Remember to plug the tank into the power pack the right way around to ensure that the DNA migrates in the correct direction—if you do not, when you return to analyze your gel, there will most

TABLE 3.3 Recipes to Prepare Loading Dye

Loading Dye 6× Concentration

Bromophenol blue—0.25% (w/v)
Sucrose—40% w/v (water)

Bromophenol blue—0.25% (w/v)
Xylene cyanol FF—0.25% (v/v)
Sucrose—40% w/v (water)

Bromophenol blue—0.25% (w/v)
Xylene cyanol FF—0.25% (v/v)
Glycerol—30% w/v (water)

Orange G—0.025% (w/v)
Xylene cyanol FF—0.25% (v/v)
Sucrose—40% (w/v) water

NB. For each of the above, sucrose (40%), glucose (30%), and Ficoll (15%) can be used interchangeably.

FIGURE 3.4   Identifying the wells and filling them with the prepared samples.

likely be nothing left on it! You can increase the voltage somewhat but be aware that increased voltage above 5 V/cm will decrease resolution of the DNA and will also create heat. Bearing in mind the size of the DNA you are analyzing and the positions that the dyes of the buffers run to, stop the electrophoresis once the samples are fully resolved. As an example, for a 500 bp DNA fragment on a 0.8% gel, using and Orange G/Xylene cyanol loading buffer, stop the electrophoresis once the Orange G has migrated just over two-thirds of the way through the gel.

3.4 STAINING AND VISUALIZING DNA

There are two ways in which you can stain the DNA within your gel, you can either include the DNA dye in the gel when you are making it, or you can stain the gel after you have finished running it. There are advantages and disadvantages to both. If you are using a DNA-intercalating dye such as ethidium bromide and you add it into the gel, it will retard the mobility of the DNA fragment. Additionally, the ethidium bromide in the gel will move in the opposite direction to the DNA and so when visualized the gel will not be evenly stained. However, staining this way is quick, as you visualize the gel immediately after it has finished running (Section 3.6). If you want to add stains into your gel, wait until it has cooled to 55°C in the water bath then add the stain to give final concentration as recommended by the manufacturer. Mix thoroughly to ensure it is evenly distributed throughout the gel.

If you do not add DNA stains when casting the gel, then you need to add it to the buffer in your gel tank once you have finished running the gel. In this case, you still need to ensure an appropriate final concentration (see manu- facturers guidelines) and leave the gel to stain for 30—45 min. Do not pipette stains directly over the gel when adding it to the tank as that can lead to bright spots on the gel, instead pipette stains into the ends of the tank then gently rock the tank to ensure it is distributed within the buffer. After the 30—45 min is up, you will need to dispose of the buffer safely and rinse the gel in water for 15 min to reduce background staining.

The most commonly used DNA satin is ethidium bromide, which is also a well-known carcinogen. If you are using ethidium bromide remember to risk assess it properly and handle using gloves. Bear in mind that you will contaminate anything that comes into contact with ethidium bromide con- taining gels or buffers, so work in a designated area and dispose of gels or buffers by decontamination. Charcoal is often used as a decontaminant, and it is a good idea to prepare a specific decontamination jar containing charcoal (charcoal "tea-bags" can be purchased from a number of suppliers for this use).

Other dyes such as SYBRsafe, Gelred, and GelGreen are now available as alternatives to ethidium bromide and all come with manufacturer's instructions advising quantity and length of staining; these sometimes require a different filter for detection and visualization. To visualize any ethidium bromide gel, you simply place it into a UV transilluminator (Section 3.6) and turn on the UV light which, if there is DNA present it will illuminate the bands.

3.5 WHICH MOLECULAR WEIGHT MARKER IS BEST FOR YOU?

In order to make sense of what you see on your gel, and to verify that the PCR product you have is the one you were after, you always need to run a molecular weight marker alongside the samples you are analyzing. Molecular weight markers are commercially available premixed fragments of DNA of known sizes (and often known quantity) that are used as a comparison to your sample. The idea is that you have already predicted the expected size of your PCR product (see Chapter 2), and by comparison to the given molecular weights of the marker, you can verify whether the PCR product you have is the correct size and therefore likely to be the product you hoped for.

When selecting you molecular weight marker, there is no "best" supplier or manufacturer to choose, all offer both broad and narrow range markers, usu- ally premixed with a loading dye, in a ready to use format. If they are not premixed, markers will ordinarily come with instructions telling you how to prepare them. The main point to consider is the predicted size of your PCR product—once you have worked out what that is, and then choose a molecular weight marker that comprises a series of fragments that cover the size range that your product falls into.

If you intend to quantify your PCR product, then choose a quantitative molecular weight marker. These have specific amounts of DNA per "band" of the marker, and you can use this information to estimate or calculate in a semiquantitative manner, the amount of DNA you have in your PCR product. This is particularly useful for downstream applications such as cloning, or if you are using end point reverse-transcriptase PCR (see Chapters 4 and 5) to analyze basic gene expression. DNA concentrations are usually given in ng per µL quantities, and the manufacturer's instructions will tell you how much you need to load onto the gel for use as a quantitative measure—it is important that you follow these guidelines or your quantification will not be accurate.

3.6 IMAGING YOUR GEL AND INTERPRETING WHAT IT SHOWS

As described previously, gels can be stained with ethidium bromide or any number of other "safe stains" that will bind to the DNA in the gel but not to components of the gel. Despite its associated toxicity, ethidium bromide is still the cheapest method by which gels are stained. If stained with ethidium bromide, gels are subsequently visualized using a UV transilluminator (often these have filters for other types of DNA stain). Typically, transilluminators come as part of an integrated system, and there are a number available on the market. Briefly, the gel is placed on a glass tray underneath which there is a UV lamp. Most gel imaging systems are enclosed to ensure that the user is not exposed to the UV light (Fig. 3.5).

When viewing the gel, you will expect the gel to appear black and the bands to show up white. You should have noted down which samples were in which lane, usually the ladder is placed in the first and last lanes of the gel. If your gel has resolved well, you will see the molecular weight marker as a series of bands running down one side of the gel, and your PCR product as a discrete band in the other lanes (Fig. 3.6).

Compare the molecular weight marker on the gel to the picture provided with the manufacturer's instructions (and which has information about the sizes/weights of the marker) and work out which bands correspond to which weight/size. Then look at your PCR product, does it appear to have resolved to the appropriate molecular weight according to what you originally predicted? If so, good, it looks as though your PCR has worked and has given you the correct product. You can never be completely certain that this is the case, to have total clarity about whether your product is the right thing you would have to sequence it, but usually the appropriate molecular weight is sufficient. Refer to Chapter 2 for problems and troubleshooting. If you see faint bands in your sample that have run significantly lower that the smallest molecular weight of your marker, these are likely to be primer dimers. If you can only see primer dimers and no product, you will need to consider titrating the primers to ensure that they do not dimerize but instead anneal to the template DNA.

FIGURE 3.5 Gel imaging equipment showing an integrated transilluminator and gel tray for use with UV light.

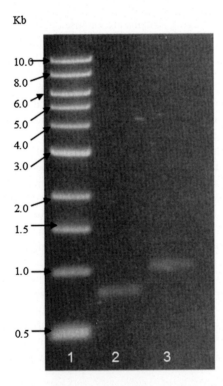

FIGURE 3.6 Gel showing a resolved molecular weight marker (lane 1) and two different sized PCR products (lanes 2 and 3). The sizes of the molecular weight marker have been indicated.

3.7 QUANTIFYING DNA/PCR PRODUCT USING ELECTROPHORESIS

There are several ways in which you can quantify your PCR product, with varying accuracy, which depends on what your downstream application is. Knowing how much (in volume) of your product you added to the gel and the amount of DNA in each of the bands of the molecular weight marker, enables you to make a reasonable estimation based on the "brightness" of the band, just by eye (the brighter the band, the higher the concentration of DNA). This is useful mainly for verifying that you have a PCR that produces a good or bad yield that you may need to optimize later.

Alternatively, you can use densitometry which is a semiquantitative method for determining DNA yield. To do this, you need to have used a quantitative molecular weight marker and you will also require access to software that enables you to undertake densitometry; with most modern transilluminators this type of software comes as standard. To quantify the PCR product by densitometry, you will use the transilluminator software (follow the manufacturer's specific instructions) to first identify the bands present on the gel. Most programs autodetect bands, but will pick up artifacts, so you may need to amend these. Focus first on the molecular weight marker, you may need to input the amount of DNA in each band (see the manufacturer's guidelines) so that you essentially tell the software what quantity of DNA the brightness of the band equates to. Some programs will do this using "brightness" alone and not actual quantities of data; this gives relative not actual quantities and therefore has limited use. Having inputted the information, next select the band that corresponds to you PCR product and use the software to calculate the density and/or the amount of DNA in the band (Fig. 3.7A and B; using a molecular weight marker as an example).

Depending on the software you have, you may only be able to acquire densitometric readings and not actual DNA quantities, in which case, using the information about your quantitative molecular weight marker, and knowing the volumes of each sample you loaded onto the gel, you will have to convert between the densitometry readings and actual DNA concentrations manually.

By far the quickest and easiest way to quantify DNA is by using spectrophotometry, although this is the best for analysis template DNA or RNA (see Chapters 4 and 5) rather than PCR products. Accurate quantification of PCR products that can be used to determine gene expression or detection thresholds can be done using quantitative-PCR, which will be discussed in Chapters 4 and 5. A number of systems are currently available to quantify DNA or RNA, including the Nanodrop or BMG Labtech LVis plate. These typically require 1 μL of DNA/RNA pipetted into a single well, and is read at a wavelength of 260 and 280 nm, both of which lie within the UV spectrum.

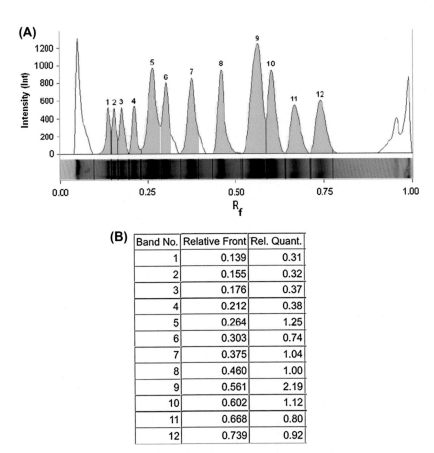

FIGURE 3.7   (A) Quantifying the PCR product by brightness. (B) Calculating the density and amount of DNA.

Such spectrometric systems provide accurate quantification of DNA/RNA as well as an indication of purity determined by the 260/280 ratio. The 260/280 ratio indicates the ratio of the readings taken at both 260 and 280 nm (usually spectrophotometric systems do this automatically) and for a good level of purity, for DNA the ratio should be 0.8 and for RNA it should be 2. Various factors can result in a low ratio, including EDTA, phenol, or proteins.

To calculate the concentration of DNA/RNA based on the 260 or 280 readings, you need to do the following calculation, bearing in mind the two points given below:

The optical density at 260 nm (OD_{260}) equals 1.0 for the following solutions:

A 50 µg/mL solution of dsDNA
A 40 µg/mL solution of RNA

Example Calculation

A sample of dsDNA was diluted 10×.

The diluted sample gave a reading of 0.25 on a spectrophotometer at OD_{260}.

dsDNA concentration = 50 µg/mL × OD_{260} × dilution factor.

dsDNA concentration = 50 µg/mL × 0.25 × 10.

dsDNA concentration = 6250 µg/mL or 0.62 mg/mL.

SUMMARY

So, having carried out your first PCR, you should now know whether it has worked and have some idea about the yield of the product. If your PCR did not work well, you should be in a position to consider the next steps to take to allow you to optimize it, be it changing the annealing temperature, adjusting the concentration of Mg^{2+}, titrating the primer pair, or preparing a new DNA template. For some applications this analysis is sufficient, but for other this is only the start of more complex optimization and gene expression studies as described in Chapters 5 and 6.

Chapter 4

Quantitative PCR: Things to Consider

ABSTRACT

This chapter will introduce the reader to fundamental concepts that require consideration prior to undertaking quantitative polymerase chain reaction (Q-PCR). It describes different types of Q-PCR as well as different applications. Advice is given regarding the type of Q-PCR machine to use, cDNA conversion, and quantification is reiterated, as well as optimizing efficiency taking into consideration such factors a primer concentration and template titration. This chapter allows readers to plan their first Q-PCR before leading them into chapter five in which the practical considerations are highlighted.

4.1 INTRODUCTION

Quantitative polymerase chain reaction (Q-PCR) is a method by which the amount of the PCR product can be determined, in real-time, and is very useful for investigating gene expression. Often abbreviated to Q-PCR, this method is sometimes also referred to as real-time PCR or depending on the application, quantitative reverse-transcriptase PCR (both of which are abbreviated to RT-PCR, which can be rather confusing). We will refer to all quantitative PCR methods as Q-PCR in this chapter, as this will distinguish between the two types of RT-PCR as necessary. Q-PCR does not rely on any downstream analysis such as electrophoresis or densitometry and is extremely versatile, enabling multiple PCR targets to be assessed simultaneously but can sometimes be a little trickier to set up than "standard" PCR; however, if you are sufficiently familiar with "standard" PCR then you are in a good position to successfully undertake Q-PCR.

4.2 WHAT ARE YOU TRYING TO ACHIEVE?

As stated previously, there are different types of Q-PCR, and it is imperative that you know exactly what it is you wish to study before you embark on this. If you are trying to detect a gene but not quantify its expression, you will use basic Q-PCR (real time) and use the data generated during the reaction to monitor the amount of PCR product over time, and the effect that parameters such as melting temperature and primer concentration have on the reaction and ultimately determine the detection threshold. If you are trying to study the expression of a gene, then you will use reverse-transcriptase PCR which means that you will have to extract RNA and convert it to cDNA prior to Q-PCR reaction. This is because the amount of RNA is related to the transcription of the gene of interest, i.e., whether it is switched on and being expressed, and using this type of RT-PCR you will obtain an indication of how highly a particular gene is expressed under a certain set of conditions.

4.3 THE IMPORTANCE OF HOUSEKEEPING

If you are studying gene expression, it is important to ensure that you have a basis for comparison for the expression of your gene(s) of interest. Usually, a housekeeping or reference gene is chosen for this purpose. Housekeeping genes are normally part of a metabolic pathway that is fundamental to an organisms' survival and whose expression is constitutive or not drastically changed in response to different environmental stimuli that may affect the expression of your gene of interest. You will need to look through literature relevant to the organism you are studying to find out what the common housekeeping genes are, and if you are lucky, you might find primer sequence is already available for these gene.

GAPDH (glyceraldehyde-3-phosphate dehydrogenase) which encodes glyceraldehyde-dehydrogenase is a commonly used housekeeping gene in most organisms ranging from man to microbe but can also be quite problematic because it has many functions besides that of the glycolytic pathway. It is crucial that the PCR reaction for your housekeeping gene is robust and well optimized; it is recommended that more than one housekeeping gene is used as a reference for normalized gene expression. Therefore, before you begin, ensure that you take the time to research housekeeping genes thoroughly to choose the most appropriate, without robust controls the data obtained will be fairly meaningless.

Two references are given below and are recommended reads when considering which reference genes to choose:

Kozera and Rapacz (2013) Reference genes in real-time PCR. *J Appl Genet* 54: 391–406.

Radonic et al. (2004) Guidance to reference gene selection for quantitative real-time PCR. *Biochem Biophy Res Comm* 313: 856–862.

Consider these as a starting point, when deciding which genes are the most appropriate housekeeping genes for your project and organism. When checking that you have chosen an appropriate housekeeping gene, assess its expression levels using the conditions that you intend to use to assess the expression of your gene of interest; the expression of the housekeeping gene should not significantly change. The purpose of the housekeeping gene is to enable you to normalize the expression of your gene of interest, essentially providing you with a baseline of gene expression against which to compare expression data and conclude whether a gene is more or less well expressed under the conditions you are investigating. It is good practice to use two housekeeping genes.

4.4 WHAT Q-PCR MACHINE SHOULD YOU USE, OR WHAT MACHINE IS AVAILABLE TO YOU?

There are many choices of PCR machine now readily available on the market and the one want to use will depends on budget and also on the amount of PCR you are going to be carrying out. There are "mini" versions that will only run 32 wells per reactions and much bigger ones that will run up to 384 wells per reactions, some will run temperature gradients to help you optimize and some would not. First of all, establish what is available to you and tailor your Q-PCR accordingly. For example, there might be specific PCR tubes or plates for a particular brand of machine or recommended Q-PCR reagents, and it is important to establish this from the outset so that you have the appropriate equipment and reagents to work with. Consider also how many channels you require; this is determined by how you intend to fluorescently label your PCR product. If you use just one label, you need just one channel; two different labels require two channels and so on. The majority of Q-PCR machines have four channels, and this is very useful for applications such as multiplex PCR.

The majority of Q-PCR machines come with specific software for analysis of results. Find out what the software is and whether it is possible to have access to it or install it on your own PC. All software is slightly different but ultimately enables you to achieve the same objective. It is recommended that you familiarize yourself with the software associated with your Q-PCR machine before you carry out your first PCR, so that you will not feel overwhelmed by the data that you obtain. Quite often, there are built in tutorials that you can work through, alternatively company website or YouTube are likely to offer a series of "How to..." videos.

Once you are familiar with the machine, the software and the reagents and consumables you need, you are in a position to consider the parameters for your first Q-PCR reaction.

4.5 EFFICIENCY PARAMETERS

The results produced from Q-PCR will not be a series of bands on an agarose gel. Rather you will be presented with a graphical representation of the accumulation of PCR product over time, usually represented as relative fluorescent units (Fig. 4.1).

The relative fluorescence units refers to the use of specific dyes that either intercalate with DNA in a nonspecific manner or are primer-specific probes that are specifically integrated into PCR product as it is synthesized (see Chapter 5). You need to determine from the data you obtain whether the reaction is optimized and can therefore be utilized for analysis. This is because you will base any judgments about the upregulation or downregulation of the gene(s) of interest, on the cycle threshold (Ct) values you generate in relation to the housekeeping genes. The Ct values are derived from the relative

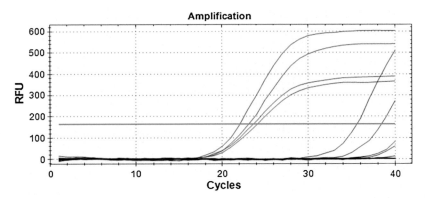

FIGURE 4.1   Accumulation of PCR product over time.

fluorescence and linked to the amount of starting material (small amount of starting cDNA = high Ct value), primer efficiency, and the amount of amplification that occurs per cycle and to ensure that you have optimized you assay you need to consider both linearity of your reaction (by creating a standard curve) and amplification efficiency (Sections 4.6 and 4.7).

We will start by considering the efficacy of the primer pair. This is the ability to anneal to the appropriate segment of DNA, without extensive formation of primer dimers, nonspecific annealing, or formation of secondary structure. All of these factors can decrease the efficiency with which the primers bind to their target DNA and therefore must be minimized. The efficiency of each set of primers must be assessed because their unique sequence confers different properties.

With a new primer pair, it is useful to prepare a primer matrix, which is a series of titrations of the primer pair as shown in table below.

		Forward primer (nM)			
		50	100	150	200
Reverse primer (nM)	50				
	100				
	150				
	200				

Following completion of the PCR reaction you will need to convert the relative fluorescence reading to Ct, using the software that is provided with the machine. For high primer efficiency low Ct values in combination with high Δ-Rn values are critical. The Δ-Rn value is the Rn value of the experimental reaction minus the Rn value of the baseline generated by the machine. The Rn value for any given reaction is the fluorescence signal normalized to the baseline.

The dissociation or melt curves for the primer matrix are also useful, these can be generated by the software provided with the instrument (refer to manufacturers' instructions) and are a plot of raw fluorescence against temperature (Fig. 4.2). If the melting temperature of specific binding events is different to that of nonspecific binding events, then it is an indication that primer dimers are not affecting the efficiency of the reaction.

4.6 LINEARITY AND AMPLIFICATION EFFICIENCY BASED IN THE TEMPLATE MATERIAL

To determine linearity, you need to prepare a serial dilution of your template material (DNA or cDNA) and then simply plot log of the starting concentration of the genetic material against the Ct values you generated from the PCR reaction. If you do not know the quantity of genetic material, you have you can plot the Ct value against the log of the dilution factor. The plot that you generate should look something like the one shown in Fig. 4.3.

As ever you should test your sample in at least triplicate when generating these curves to ensure reproducibility. The log nanograms have been converted

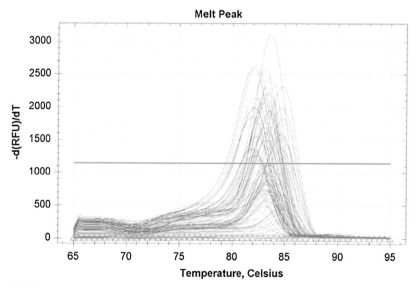

FIGURE 4.2   Melt curves generated from a primer titration matrix.

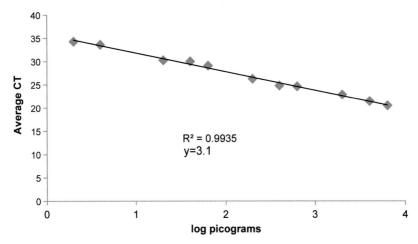

FIGURE 4.3   Ct value over the log of the dilution factor.

into picograms (just multiply by 1000) on this graph to avoid negative log numbers. If you do not want to convert to picograms, you need to incorporate the negative numbers into the equations used later on. Once you have plotted the standard curve you need to determine either the; r value (Pearson's correlation coefficient) or the R^2 value (coefficient of determination), this can be dome automatically in software such as Excel by double clicking on the trend line and selecting "display R^2 value."

The linearity of the graph should ideally be: r greater than 0.99 or R^2 greater than 0.98, to ensure accurate quantification of your samples. The values indicate how linear the data are, if you have a low value, it indicates that you have variability across the dilutions you prepared, e.g., that the amount of starting material is affecting the amplification.

4.7 AMPLIFICATION EFFICIENCY

PCR amplification efficiency is normally expressed as a percentage and indicates the rate at which your PCR product is generated. In an ideal world, you want the product to double with each round of amplification and if the efficiency value is 100% that is what you have achieved. To generate your E value, you need the slope of your standard curve. The slope is indicated by the y value (see Fig. 4.3), in this case y = 3.1.

To use this fit it into the following equation:

$E = 10^{1/slope}$

Then to convert into a percentage:

% Efficiency $= (E - 1) \times 100$.

So using the slope generated by the graph (y value) above you would get:

$$E = 10^{0.3}$$
$$= 1.99.$$

Therefore % efficiency $= (1.99-1) \times 100 = 99.5$.

This would indicate to you that during one cycle of the PCR reaction your copy number has increased by 99.5%. Obviously, the closer to 100% efficiency you can get the more robust your PCR will be but generally any figure between 90% and 105% is considered acceptable.

There are several reasons why you might get figures outside of this range including badly designed primers with secondary structure, contamination with inhibitors or more basic errors such as not pipetting your dilutions accurately. If the efficiency values do fall outside of those parameters, then it is probably time to redesign your primers.

4.8 cDNA CONVERSION AND QUANTIFICATION

The descriptions above have referred to the template material. For Q-PCR, the template material can be DNA or cDNA. If you are using cDNA, you first need to extract RNA from your samples. To do this follow the same procedure as described in chapter one but use acidic (pH 4.5) phenol:chloroform:isoamyl alcohol (25:24:1). The acidic pH means that most of the DNA remains in the solvent phase, and the RNA is in the aqueous phase. All reagents you prepare for RNA work must be treated with DEPC (diethyl polycarbonate) to remove RNAses, and it is recommended to purchase reagents and consumables specifically for RNA work that have been appropriately treated to remove RNAses. There are a number of commercially available RNA extraction kits available. It is advisable to use a commercially available kit, especially if you have never handled or extracted RNA before. This is to guarantee that all of the reagents and consumables are RNAse free and does not require you to treat reagents and consumables.

RNA is quantified by spectrophotometery as described in Chapter 3.

Having extracted RNA from your sample you now need to convert it into cDNA. The conversion to cDNA is accomplished by using the enzyme reverse transcriptase. Again, there are many kits on the market that you can buy to do this, or you can prepare the reagents individually and optimize concentrations yourself. If you choose the latter ensure that you have a designated "RNAse free" work space, consumables, and equipment. Basic reagents required are:

Reverse-transcription buffer, dNTPs, oligo(dT), or random oligonucleotide primers, reverse-transcriptase and nuclease-free water. A number of commercially available kits can be purchased to do this. The majority utilizes random primers for synthesis and come with the appropriate buffers and parameters; follow the manufacturer's guidelines for this step and these are kit specific. You will require a thermal cycler for cDNA conversion. The sooner you can remove your cDNA from the thermal cycler and use or store it, the better. For short-term storage, cDNA can be kept at $2-6°C$ for longer term, store at $-80°C$.

It is not normal to try and quantify cDNA as there can be carryover from the dNTPs and degraded RNA, in practice RNA is quantified, and it is assumed that cDNA conversion is complete. However, you need to account for the fact that you have diluted your sample in the reaction buffer during cDNA conversion.

4.9 PLANNING TO USE Q-PCR FOR MULTIPLEX

It is possible to assess the expression of multiple genes at the same time or genes of multiple origins in the same sample. This requires the addition of two or more sets of primers to your reaction and can be quite tricky to optimize since you must have multiple primer sets that work equally as well under the same conditions. When this is done, it is called multiplex PCR. Most Q-PCR machines have more than one channel; channels detect light emitted from tagged DNA or labeled primers. Q-PCR machines may have four or even more channels, the number of channels dictates how many labeled primer pairs you can use in your PCR.

For example, if you plan to undertake a PCR reaction using four differently labeled primer pairs, you will need a machine with four different channels to detect each of the four different, labeled PCR products. You will also need to select different markers for you PCR products—the easiest way to do this is by using probes that bind specifically to the double-stranded PCR product—the more PCR product that accumulates, the greater the signal from the complementary probe. Common labels for probes include DAPI (4'6-diamidino-2-phenylindole), Texas Red, and FITC (fluorescein isothiocyanate), most primer design companies will have a list of labels available for use and it is up to you to decide which to use. Make sure that each probe produces a different signal, some probes produce the same signal and so if used together will not allow you to distinguish between two different PCR products. When optimizing a multiplex PCR, the same parameters as those described throughout this chapter must be considered.

SUMMARY

Having planned your Q-PCR and given consideration to optimizing the parameters and interpreting the data and having hopefully optimized the reactions for your housekeeping genes, you are ready to undertake gene analysis, which will be described in Chapter 5.

Chapter 5

Carrying Out Q-PCR

ABSTRACT

Having introduced some of the concepts of quantitative polymerase chain reaction (Q-PCR) in the previous chapter, this chapter deals with some of the more practical considerations for setting up and running Q-PCR. It provides information about reagents, commercially available kits, and suppliers as well as suggesting some basic protocols that can be easily modified by the user. Additionally, it describes how to analyze data using the $\Delta\Delta Ct$ method and how this can be converted into fold-changes to give an indication of differential gene expression, normalized to a reference or housekeeping control.

5.1 INTRODUCTION

As mentioned in Chapter 4 when preparing your quantitative polymerase chain reaction (Q-PCR) experiments, remember that you will primarily be handling RNA so it is important that you ensure the environment in which you work is RNase free. Take the same precautions as were described in previous chapters; it is a good idea when undertaking Q-PCR, to keep a set of pipettes just for this purpose ensuring you clean them well after each use. If possible keep the RNA you are working with isolated from any DNA samplesand any extraction kits you might have been using, which often contain RNase.

5.2 REAGENTS

Reagents for Q-PCR are available form a number of suppliers (see Chapter 4, although this list is not exhaustive). The majority of Q-PCR reagents are provided as part of a kit which includes a master mix comprised of reaction buffer, dNTPs and Mg^{2+}, and you will add to this the template material and enzyme. Depending on the application, your template material might be DNA or cDNA; if the latter then remember that you will need to convert the RNA you extracted previously, into cDNA (Chapter 3).

The vast majority of Q-PCR kits utilize SYBR green (N′N′-dimethyl-N[4-[(E)-(3-methyl-1,3-benzothiazol-2-ylidine)methyl]-1-phenylquinolin-1-ium-2-yl]-N-propylpropane-1,3-diamine) as a DNA intercalating dye, although the trade name will vary depending on the manufacturer. Specific probes are available, such as TaqMan, or you can design you own primer probe combinations, for example, if you are planning a multiplex PCR. Some common fluorescent probes include DAPI (4,6-diamidino-2-phenylindole), FITC (fluorescein isothiocyanate), and Texas Red. When you order your primer pairs, you can choose the type of probe you want and it will be synthesized for you.

It is prudent when using a Q-PCR kit to follow the guidelines as recommended by the manufacturer in terms of volumes and quantities of reagent. Draw up your own "recipe list" as per basic PCR guidelines and pin in up on or near your bench for reference. Consider how may reactions you need to carry out. If just a few, then it's advisable to use PCR strips with optical lids. These come in strips of eight that can be broken into smaller strips or individual tubes if necessary (Fig. 5.1). They are provided free of DNase and RNAse and as such are ready to use. If you have a large number of reactions to prepare, you can use a plate with either optical lids or optical plate seals. Check which type of plate fits the machine you have before you purchase.

If you do not have access to a Q-PCR machine, you can still analyze gene expression in a semiquantitative manner using end point RT-PCR. To do this, it is still necessary to extract RNA and convert it to cDNA. The cDNA is used in the RT-PCR reaction which is subsequently analyzed by electrophoresis. You can determine the amount of the end product using densitometry, and this can be compared to a housekeeping gene and differential expression inferred from this. To do this, you need to generate densitometry readings as described in Chapter 4, and it is important to use a quantitative molecular weight marker

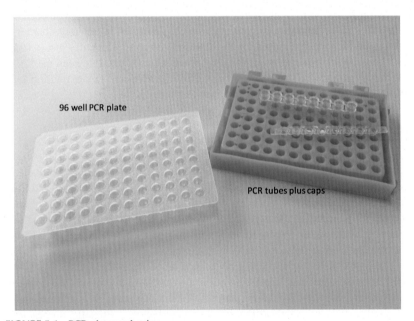

96 well PCR plate

PCR tubes plus caps

FIGURE 5.1   PCR plates and strips.

(i.e., the amount of DNA in each band is known). Using a housekeeping gene that is not differentially expressed under either test condition, calculate the density units per ng for several bands of the molecular weight marker (and then calculate an average); utilize this value to convert the densitometry readings for your unknown and/or samples into ng quantities.

Because end point RT-PCR gives no indication of the cycle threshold, it is only semiquantitative. A number of companies sell cDNA conversion and end point RT-PCR kits and some combine the two in a one-step end point RT-PCR kit. Which you use is up to you, but it is advised that you follow the recommended instructions provided by the manufacturer. Densitometric analysis can be carried out as described in Chapter 3.

5.3 BASIC PROTOCOLS

Using the guidelines in Chapter 4, you will have prepared cDNA (unless you intend to use DNA), which you will use for your Q-PCR reaction, whether you are using end point analysis or not. You will have quantified the cDNA following conversion and decided on how much to add to the reactions. The protocol you use will be similar to the one you originally optimized for your basic PCR in terms of annealing temperature and the concentration of the reagents; this will be your start point but you might still need to optimize further as described in Chapter 4.

So essentially at this point, you need to prepare your reactions as recommended by the manufacturer and program the Q-PCR machine with the parameters that are appropriate for your reaction. If you have optimized "standard" PCR parameters prior to moving to Q-PCR, then these will already be optimized. Remember that additional optimization as described in Chapter 4 might be necessary. Next will follow the interpretation of the data you obtain post-Q-PCR.

Once your PCR reaction is complete, the Q-PCR machine software will produce a graphical representation of the accumulation of PCR product in each of your samples (Fig. 5.2). This example shows a completed run for two housekeeping genes, and eight test conditions, all performed in triplicate. The curves that are just beginning at cycle 38 are the negative control and indicate that the PCR is robust.

The negative control should show no or negligible amplification within the reaction parameters. If is not the case, then it suggests your negative control is contaminated. For your other samples, you need to assess first, how much PCR product is produced, and the point at which the optimum amount is produced, allowing you to determine a threshold for detection that you will use for all of your subsequent experiments. This will be determined by the concentration of your template material, a specific primer ratio and a specific number of cycles (see Chapter 4). In Fig. 5.2, the threshold is indicated by the green horizontal line. There will be guidelines provided with the machine you use to help you to set the threshold, refer to these instructions, and use

Amplification

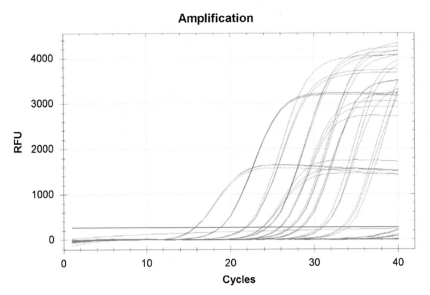

FIGURE 5.2   Accumulation of PCR plotted by sample.

the software provided in combination with your optimization data to establish the threshold for detection. Once established, this threshold will be used for future experiments.

5.4 ABSOLUTE QUANTIFICATION

The type of analysis and quantification you choose depends upon your initial goal. Absolute quantification enables you to determine expression levels for a gene in absolute numbers of copies. In order to do this, you require a set of standards of known concentrations from which the values for your unknown targets can be extrapolated. To achieve this, a standard curve must be generated (see Chapter 4), and the precise amount of template used to generate the curve must be known (this can be obtained by spectrometry as described in Chapters 3 and 4).

The guidelines below are critical for generating a standard curve for absolute quantification:

The DNA, mRNA, or cDNA must be of high quality and "clean" (see Chapter 4).

Accurate pipetting is required when preparing the standards because they are diluted over several orders of magnitude; ensure that the standards are diluted to a range that ensures you to achieve a concentration similar to that of the unknown target(s). If you do not, then it will not be possible to extrapolate from the standard curve.

Prepare the standards and the samples fresh before use to avoid degradation that will interfere with the results and therefore the standard curve. If you have pre-prepared standards store them at $-80°C$, and only use them once as repeated freeze-thawing results in degradation. Remember that if you are quantifying RNA, DNA standards should not be used. The absolute quantities of the standards must be determined by an independent means, quantification by spectrophotometry, and determination of absolute quantities based on A260 readings is recommended (see Chapter 4). When preparing reactions for the standard and unknown, both are usually simultaneously run and analyzed. So your reaction with be comprised of a set of standards, a positive control, a negative control, and your unknown samples. Replicates are critical, and it is advised to use at least three replicates per sample.

As described in Chapter 4, once the reaction is completed you will need to generate a standard curve based on the Ct values obtained for your standards. As previously described, you can determine from this the efficiency of the reaction as well as absolute quantities. Having plotted your standard curve, you can calculate the absolute quantity of your unknown sample as described in Chapter 4.

5.5 RELATIVE QUANTIFICATION

Relative quantification is a little different but is still reliant on the production of a standard curve, or curves. This type of quantification is used to determine fold changes in expression between two samples and is normalized to a housekeeping gene, utilizing the same samples. Again, it is critical to set up replicate reactions to ensure that you have robust, consistent data.

When preparing Q-PCR for relative quantification, you will have a reference sample which you will use to "normalize" or calibrate the information you acquire about the unknown or test samples. This is often a housekeeping gene for which expression remains unaltered irrespective of the conditions of your experiment. Therefore, the results for your unknown or test samples will be expressed as, for example, "under aerobic conditions sample A had a 12-fold greater expression that under anaerobic conditions."

To establish fold-changes, you need to use the comparative Ct (or $\Delta\Delta$Ct) method as described in the following paragraphs.

The $\Delta\Delta$Ct method compares the Ct value of one target gene to another (the housekeeping or reference gene). It does not require the generation of a standard curve, but to be valid, the efficiency of the reference and unknown and/or sample reactions must be approximately equal. The following bullet points and table sets out how to calculate $\Delta\Delta$Ct, but the majority of software packages provided with Q-PCR machines will do this and the information can be exported to programs such as Excel.

- Calculate the average Ct values for your reference and/or housekeeping gene and you unknown and/or sample(s)
- Calculate the difference between the Ct value for the unknown and/or sample and the Ct for the reference and/or housekeeping gene (this gives the ΔΔCt value)
- Use the formula $\log_2\Delta\Delta Ct$ to calculate the fold change

ΔCt sample	ΔCt reference	ΔΔCt	Fold change $(\text{Log}_2\Delta\Delta Ct)$
2.45	0.23	2.22	4.44

Alternatively, you can use the standard curve method. This generally requires the least amount of optimization and means you can run the reference and/or housekeeping reaction and unknown and/or sample reactions separately. However, it is important that you select a reference and/or housekeeping gene that is consistently expressed and whose expression is not altered by the experimental test conditions; it is advisable to use at least two reference and/or housekeeping genes if using the standard curve method.

Standard curves must be generated for the reference and/or housekeeping gene and the unknown and/or sample; therefore, if you opt for this method of analysis your dilutions must be very accurate, and as usual, it is imperative to use replicates. You will also have to run a validation experiment demonstrating that the amplification efficiencies of the reference or housekeeping green and unknown or sample are approximately equal.

To achieve the required standard curves, a series of dilutions should be prepared; at least five which can be 2-, 5-, or 10-fold. Prepare separate reactions for each dilution. As described in Chapter 4, use the Ct values obtained from the reactions to plot a curve of Ct against dilution factor, ensuring that the correlation is 0.99 or higher. Having established standard curves for both the reference or housekeeping gene and the unknown or sample, the relative expression level of the unknown and/or sample assessed under different conditions, can be extrapolated.

To convert the relative expression changes into fold changes, the extrapolated values must be normalized compared to the standard curve of a "calibrator" sample; a calibrator must be selected for both the reference or housekeeping gene and the unknown or sample. Ordinarily, the calibrator is the "untreated" sample (for both reference or housekeeper and unknown). For example, if you are asking whether iron restriction results in increased expression of the Ferric Uptake Regulator gene in *Escherichia coli*, you would be assessing two conditions—"untreated" and "treated with an iron-chelator," i.e., iron restriction; the calibrator in this case is the former.

Having obtained Ct values for both the reference or housekeeping gene and the unknown or sample you can utilize the $\Delta\Delta Ct$ method. To begin with you need to normalize, the Ct of the target gene to the reference gene as shown below:

$$\Delta Ct \ (\text{for test conditions}) = Ct \ \text{unknown or sample}$$
$$- \ Ct \ \text{reference or housekeeping gene.}$$

$$\Delta Ct \ (\text{for "untreated" i.e., calibrator}) = Ct \ \text{unknown or sample}$$
$$- \ Ct \ \text{reference or housekeeping gene.}$$

Next, you need to normalize the ΔCt of the unknown and/or sample to the ΔCt of the reference and/or housekeeping gene; this generates the $\Delta\Delta Ct$ value:

$$\Delta\Delta Ct = \Delta Ct \ \text{for the test conditions}$$
$$- \ \Delta Ct \ \text{for the "untreated" and/or calibrator conditions.}$$

This value can be used to calculate the fold change in expression using $\log_2\Delta\Delta Ct$, as previously described.

Please note that for this method it is necessary to produce a standard curve based on fold-dilution for each gene you are studying and the reference and/or housekeeping gene.

Having mastered both the absolute and relative quantification techniques described above, it is prudent to prepare a "master" spreadsheet containing the relevant formulae that comprise the fold-change calculation. This way you can transfer the Ct values for your reactions directly into the spreadsheet and the fold changes will be automatically calculated, saving you a lot of time.

If you are lucky, the software provided with your PCR machine will enable you to generate standard curves and some will also enable you to calculate fold-changes; refer to the manuals that are provided with your machines for details on specific software. However, it is good to know exactly how these values are established so that gene expression analysis does not become a "black-box" that you struggle to comprehend.

SUMMARY

Having familiarized yourself with the contents of both Chapter 4 and this chapter you should be in a position to optimize and run Q-PCR reactions to assess gene expression. The first time you do this it can seem like a long and arduous process, and preparation is a key. However, having carried out Q-PCR and downstream analysis once, then subsequent reactions seem easier—it is a case of practice and getting used to the methods of analysis.

Chapter 6

Using PCR for Cloning and Protein Expression

ABSTRACT

This chapter will describe how polymerase chain reaction amplified genes or gene fragments can be used in cloning experiments for a number of applications including complementation, functional analysis of proteins, or proteins overexpression prior to purification by chromatography. The reader will be able to utilize the skills learned in Chapters 1—3 and apply them to the techniques described here.

6.1 INTRODUCTION

In addition to gene expression analysis, polymerase chain reaction (PCR) can be used for a number of additional applications. This chapter will explain how PCR can be used to clone genes or gene fragment into plasmid vectors that can be subsequently used for protein expression studies or for the overexpression and subsequent purification of proteins (although protein purification techniques will not be described here). Having read through Chapters 1—3 you will already have the basic skills required to optimize and undertake the PCR reactions needed for these applications. Therefore, the information in this chapter will build on these skills to allow you to utilize PCR as a molecular tool. This chapter assumes that you are using a prokaryotic system for analysis.

6.2 ENZYMES FOR CLONING

Taq polymerase remains to be the most ubiquitously used polymerase for PCR. This enzyme has a high rate of dNTP incorporation but does not have any proof-reading activity. This means that Taq is prone to introducing base-pair errors during PCR amplification. If the PCR product is to be used for functional protein analysis, then this can be problematic. Single base pair changes have the potential to change the amino acid sequence which could in turn impact upon the protein product and its functionality. However, Taq does have a very useful characteristic that is exploited for cloning. Namely that it incorporates a single adenosine base at either end of the PCR product. This in

61

essence creates "overhangs" that will readily bond with thymidine nucleotides by complementary base pairing—how this can be used for cloning will be described later in this chapter.

To overcome the problem associated with Taq-introduced errors during PCR, alternative polymerase must be used. Pfu is a polymerase with $3'$-$5'$ exonuclease activity meaning that as the PCR product is synthesized it checks which nucleotides have been incorporated and can work backwards to remove incorrect bases and replace them with the correct ones. Almost all suppliers of polymerase enzymes for PCR will produce a version of a so-called "proof-reading" enzyme, check the product descriptions before you buy. The drawbacks of using "pure" preparations of proofreading polymerase are that they do not incorporate an adenosine base at the ends of the PCR product. This can be overcome by using a Taq/Pfu mixture which many companies produce—the merits of using these enzymes mixtures will be described later on in this chapter.

So, it is highly likely that you will need to use a high-fidelity enzyme, and it does not really matter which one you choose. The method of cloning you wish to utilize will ultimately dictate this decision.

6.3 "CLEANING" PCR PRODUCTS FOR CLONING

There are a number of commercially available cloning kits that provide the enzymes mixtures and plasmid vectors necessary for cloning. These usually come with preoptimized protocols, and it is advisable to follow those if you are using a kit. Otherwise, there are a number of things that you need to do before you can introduce your PCR product into a plasmid vector.

First, it is imperative that you establish that you have a single, pure PCR product, i.e., a single band when analyzed by electrophoresis (see Chapter 3). Next, you need to "clean" the PCR product to remove residual enzyme and salts from the PCR reaction that could potentially interfere with ligation. Again, there are a number of PCR clean-up kits that are commercially available. Many of these use a spin column system and a microcentrifuge. These work on the basis of binding DNA (the PCR product) to a membrane and washing away impurities by addition of a series of pre—prepared buffers and ethanol. Clean DNA is then eluted from the membrane in molecular grade water.

If you do not use a PCR clean-up kit, you can always purify the PCR product by ethanol precipitation as described in the following list (intended for microcentrifuge tubes):

1. Add 0.1x volume of 3M sodium acetate (pH 5.2) or $0.5\times$ volume of 5M ammonium acetate
2. Next, add $2.5\times$ volume of 100% ethanol (ice cold)

3. Leave the preparation in the freezer for 30 min (or overnight) to allow the DNA to precipitate

4. Centrifuge preparation at maximum speed for 30 min at 4°C—the DNA pellet might not be visible, so mark on the tube so you know which way around it is and therefore know where the DNA pellet will be post-centrifugation.

5. Decant the supernatant and air-dry the pellet for 15—20 min on your bench

6. Resuspend the pellet in an appropriate quantity of water or TE buffer—if you need to concentrate the amount of DNA you have at this point you can do so by reducing the amount of buffer, for example, if you prepared a 50 µL PCR reaction, you can dissolve the DNA in 25 µL buffer or water and will therefore have prepared a sample of higher concentration.

If you have a troublesome PCR that produces more than one product, as observed by producing more than one band by electrophoresis you can always extract the PCR product you need for cloning from the agarose gel. If you have to do this, ensure that you can confidently identify the band that corresponds to your gene and/or PCR product of interest. Again, numerous suppliers of molecular reagents produce Gel-Extraction Kits, which use a series of buffers that will first dissolve the agarose and enable the DNA therein to be extracted, usually using a spin column (as described above). The first stage of gel extraction will require you to cut the band of interest out of the agarose gel using a scalpel. You will need to identify the position of the band of interest using a UV transilluminator and then cut out the band ensuring that you cut as closely to the band as possible so that you do not have too much excess agarose gel. When you have cut the band out, check the gel again by looking at it under UV light—you should have a small block of agarose that glows (because the PCR product is in it) and an empty space in your gel with no residual DNA left around it. This can be tricky to do so it is a good idea to practice a bit first.

Irrespective of the way in which you clean up your PCR product it is advisable to quantify it postpurification because each methodology will invariably lead to some loss of yield. When you begin cloning, you will need to know that concentration of PCR product so that you can calculate PCR product:vector cloning ratios.

6.4 DIRECTIONAL CLONING

The direction that the PCR product is incorporated into the plasmid vector is imperative for appropriate protein expression. Often plasmids for expression are designed with specific promoters to drive protein expression, or they might include "tags" for protein purification that are incorporated at the 5' or 3' end of the cloned DNA fragment. If you are using the promoter region that is native to your gene of interest, then the direction is less critical.

It is therefore imperative that you have already identified the plasmid vector you intend to use and have designed primers to allow you to directionally clone the PCR product. There are a number of ways to do this. For simple directional cloning using you can incorporate restriction endonuclease sites into your PCR product. To do this, you need to first identify which restriction enzyme sites are available in the plasmid you are cloning into. Commercially available plasmids are designed to have a number of restriction sites located within the "cloning site" and information can be obtained from the companies from which you buy the plasmid.

You will need to identify two different restriction enzyme sites. Design your primers as described in Chapter 2 and add the sequence for the restriction site you have selected to the 5' region of the primer (Fig. 6.1). These will be incorporated into you PCR product during the PCR reaction.

When you are ready to clone the PCR product into your plasmid vector of choice, you will need to digest both with the restriction enzymes you selected. Most companies that produce restriction enzymes will provide a universal enzyme buffer so that you can digest with both enzymes simultaneously, but if not you will need to do to consecutive digestions with a clean-up step (see Section 6.3) in between.

Digestion of plasmid vectors is often more successful than that of PCR products—this is because there is simply more DNA for restriction enzymes to "attach" to. To maximize the digestion of the PCR product, you can include an intermediate cloning step using a T-tailed vector. If you wish to do this, you will need to use a Taq/proof-reading enzyme mix as described in Section 6.2. Ligate the PCR product into the T-tailed vector following the manufacturer's instructions (Fig. 6.2). The cloning will not be directional but that does not matter at this stage. You will need to propagate the vector prior to digestion and purify the digested fragment by gel extraction (see Section 6.5). The clean digested fragment will be used in subsequent ligation reactions.

If you are adding a specific tag to your PCR product, then sequence to enable you to do this has to be incorporated into the primer as described above for restriction endonuclease sites. Often this sequence is not the entire tag sequence as it were but is sequence that it complementary to the cloning site in the vector you are using. If you are using this type of vector system, it is important to follow the manufacturer's instructions for primer design and subsequent ligation. This type of cloning kit is often called a one-step

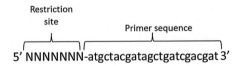

FIGURE 6.1   Restriction site and primer sequence.

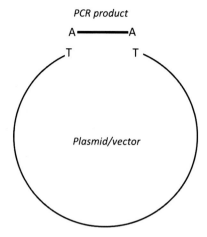

FIGURE 6.2   T-tailed vector.

directional cloning kit and has a very high rate of success; traditional directional cloning using restriction enzymes can take longer to optimize and this will be addressed in the following section.

6.5 LIGATION OF PCR PRODUCT AND VECTOR

The vector use to ligate your PCR product into will ordinarily be of commercial origin which means you will have all of the appropriate information you need with regards to the cloning site, size, and concentration. You will need this information to prepare the ligation reaction. You will also need to know the quantity of PCR product you have to ligate into the vector.

If you are using a T-tailed vector system, use the instructions provided by the manufacturer to guide you. It is recommended to use a range of vector:PCR product concentrations, so begin with a range of 1:3, 1:1, and 3:1 (vector:PCR product). The ratio is calculated as described in the following equation:

ng PCR product for 1: 1ratio = kb PCR product/kb vector × ng vector.

Adjust the values accordingly for different ratios.

This formula can be used for any vector/PCR product combinations and is not restricted to T-tailed vectors.

Once you have completed the ligation you will need to propagate the plasmid (see Section 6.7). After which you will digest the fragment out of the T-tailed vector, resolve the digested construct by electrophoresis, clean it by gel extraction (you will cut out the band corresponding to the size of the cloned insert) and prepare a second directional ligation reaction.

To directionally clone the PCR fragment, you will first prepare the fragment as described above and also the vector by digestion with the same two restriction enzymes. Once the PCR product and vector have been digested, it is important that they are used immediately and kept on ice during setup of the ligation reaction as they will be prone to degradation by nucleases.

For the ligation reaction, you will need to use a ligase enzyme. These are available commercially and it does not matter which manufacturer you use; however, it is advisable to follow the manufacturer's instructions with regards to reaction volumes, incubation times, and appropriate reaction buffers. Start by calculating how much vector and inset you will need for the cloning ratios you have selected to use. Using the information provided by the manufacturer, prepare for yourself and ingredient list for each of the different ratios and pin it up near your bench for reference when setting up the reaction. Usually the ligation reaction will include the following:

PCR product
Vector
Ligase enzyme
Reaction buffer
Water (to make the reaction up to the recommended volume)

Ligase reactions are usually carried out at 16°C for 2–6 h but can be incubated at 4°C overnight. A number of rapid ligation kits are available that enable the ligation reaction to be completed in much less time—it is up to you which you choose to use.

If you are using a one-step directional cloning kit, the likelihood is that ligase enzyme and reaction parameters will be provided as part of the kit, and it is imperative that you follow the instructions provided by the manufacturer with regard to the reaction ingredients, parameters, and suggested cloning ratios.

Once the ligation reaction is complete, the new plasmid construct must be analyzed to determine whether the ligation has been successful. To do this, you need to first propagate the plasmid construct, this is usually done by transforming it into a bacterium such as *Escherichia coli* as described in the next section.

6.6 TRANSFORMING AND ANALYZING THE PLASMID CONSTRUCT

Plasmid constructs need to be propagated prior to analysis to see whether the PCR product has been successfully inserted or not. The analysis described below can be used for any type of ligation, T-tailed, directional, or one-step cloning because it is based on digestion of the plasmid construct, and all commercially available vectors will have restriction enzymes sites located either within the cloning site or elsewhere on the vector.

If you are using a commercially available kit, it will most likely be provided with competent cells and instructions describing how to transform them; if this is the case then it is recommended that you follow the manufacturer's instructions. If not there are a number of ways in which you can prepare and transform bacterial cells. The most commonly used methods use heat shock or electroporation. For both, the bacteria must be specially prepared to make them competent (not all bacteria are naturally competent).

Chemically competent bacteria are prepared as described below. But before you begin ensure you know which strain to transform, for checking plasmid constructs something like DH5-α or JM101 are useful. You also need to know what you will use to select for the transformants, for example, what antibiotic selection to use or whether you can utilize blue-white selection.

Preparing Chemically Competent Cells for Transformation by Heat Shock

1. Culture the bacteria to an optical density of 0.2–0.5 (A_{650})—use 2×50 mL volumes
2. Centrifuge the culture at 4000 rpm for 15 min at 4°C
3. Decant the supernatant and resuspend the pellet in 50 mL ice-cold 0.1 M $CaCl_2$
4. Incubate the cell suspension on ice for 30 min and then centrifuge as described in (2)
5. Decant the supernatant and resuspend the pellet in 5 mL of an ice-cold solution of $CaCl_2$ in 15% glycerol
6. Aliquot 50 μL portions into sterile microcentrifuge tubes, use immediately, or store at −80°C

Transforming Chemically Competent Cells by Heat Shock

1. Keep the tubes of cells on ice to thaw
2. Prepare a water bath or heat block at 42°C
3. Add 3–5 μL of plasmid construct to the cells and mix gently with the pipette tip
4. Continue to incubate the cells on ice for a further 5 min
5. Transfer the cells to the water bath or hot block for 45–60 s and then place them back onto the ice
6. Incubate the tubes at 37°C for 20–30 min
7. Spread the cells onto an agar plate with appropriate selective ingredient, e.g., antibiotics and incubate at 37°C overnight.

Preparing Cells for Electrocompetent Cells

1. Culture the bacteria to an optical density of 0.5 in super optimal broth (SOB) media (2% tryptone, 0.5% yeast extract, 10-mM NaCl, 2.5 mM KCl, 10 mM $MgCl_2$, and 10 mM $MgSO_4$, sterilized)

2. Centrifuge cultures (2 × 50 mL) at 4000 rpm for 15 min at 4°C and pour away the supernatant
3. Keeping the cells on ice, resuspend the pellet in 50 mL ice-cold glycerol, and centrifuge as described above
4. Resuspend the pellet in 50 mL ice-cold glycerol and centrifuge as described above
5. Resuspend the pellet in the residual glycerol and aliquot 50 μL portions into sterile microcentrifuge tubes.
6. Use immediately or store at −80°C

Transformation by Electroporation

1. Set the electroporator to 1.7−2.5 kv (the optimal kv will depend on the strain of bacteria you are using and may require some optimization), 200 Ω and 25 μF.
2. Thaw electrocompetent cells on ice
3. Chill the electroporation cuvettes on ice
4. Prewarm the recovery media (SOB + 20 mM glucose)
5. Add 3−5 μL of plasmid construct to the cells and gently mix with the pipette tip
6. Transfer the cells into the electroporation cuvette and wipe any water from the outside of the cuvette
7. Place the cuvette into the electroporation module and press the pulse button (follow the instructions for the machine you are using, these will vary depending on the manufacturer)
8. Transfer the cuvette back onto ice and immediately add recovery media to a total volume of 1 mL
9. Transfer the contents of the cuvette to a sterile microcentrifuge tube and incubate for 20−30 min at 37°C
10. Plate out and incubate the cells as described for chemically competent transformation

NB. For both types of transformation, you can either plate out the cells directly or prepare dilutions, i.e., 10-fold, 100-fold prior to plating.

After overnight incubation, you are in a position to screen the transformants to see if they contain the plasmid construct.

Prepare a fresh agar plate (containing appropriate selective ingredients) and draw a grid on the back of it—number the grid. Select colonies from the plate of transformants and inoculate them into the segments of the grid. Using a sterile yellow pipette tip, draw a diagonal line in the square of the grid, ensuring that none of the lines touch each other. Incubate the plate at 37°C overnight—these are the transformants you will screen.

There are a number of ways to screen the transformants. You can use PCR as an initial screen. To do this, you can use either the primers you used originally to prepare the PCR product for cloning, or you can use a new set of primers that span the cloning site and your insert.

You can prepare the screening PCR at the same time as your grid plate. To do this, setup as many PCR tubes as you have squares on the agar grid plate—number both tubes and plates accordingly. Add 50 μL PCR grade water to the tubes. Touch as sterile yellow tip to the colony you wish to screen (from the plate of transformants) and "stir" it into the first PCR tube and then draw a horizontal line on the corresponding grid square (there will be enough residual bacteria to grow).

Once you have worked your way through all of the selected transformants, incubate the grid plate as described above, and boil the contents of the PCR tubes for 10 min. You can setup a program on the PCR machine to do this. Prepare the PCR reagents as a master mix and aliquot into fresh PCR tubes. Add 5 μL of the "boilate" to the individual PCR tubes (remembering to number them accordingly) and run the PCR reaction. Analyze the reaction by electrophoresis—if the PCR product has been successfully ligated into the vector you will see a PCR product of the appropriate molecular weight. The gel might look a bit messy with multiple bands, but you should be able to identify your fragment of interest.

Make a note of which transformants contained the insert by PCR screen. Next, prepare 5—10 mL culture of each transformant and culture them overnight (ensuring to use the right selection to maintain the plasmid). Then extract the plasmid and digest it with the same enzymes you used to directionally clone the fragment into the plasmid in the first place. This will "release" the fragment. Analyze the digested plasmid construct by electrophoresis of the insert is present you will see a band of the appropriate molecular weight of the original PCR product (Fig. 6.3).

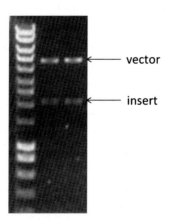

FIGURE 6.3 Confirming the presence of the insert.

Alternatively, you could digest the construct using a restriction enzyme that cuts only within the insert. Or you can choose one enzyme that cuts in two or more places in the plasmid construct and look at the difference in size for fragment containing your insert. If you are using this method to screen PCR products that have been cloned into a T-tailed vector, once you have verified the presence of the insert, you will need to digest the entire plasmid preparation (to "release" the insert), electrophorese it and extract and clean the digested insert (using a gel extraction kit, as previously described), and then ligate it into your expression vector, essentially as described in the previous sections of this chapter.

Once you have identified "successful" plasmid constructs, prepare a stock, and store it at $-80°C$. You are now ready to use the construct in expression experiments.

6.7 SEQUENCING PLASMID CONSTRUCTS

By this stage, you should be reasonably confident that the gene of interest is successfully cloned into the plasmid and that you have a functional plasmid construct. However, it is always good practice to check that all is well by sequencing the cloned gene. This will verify that the sequence is correct. You can use any sequencing service you prefer and each will provide you with a preferred means of preparing your plasmid construct and shipping it. Sequencing can be done using your own primers or those of the sequencing company. Check the details for your vector for which primers to use, often T7 and SP6 are utilized, and these prime to the plasmid itself rather than your insert.

You will receive the sequence data in an electronic format, both as a FASTA (FAST-All) sequence and as a trace. When you see the trace, there will be four different colored lines corresponding to each of the four DNA base pairs. The series of peaks indicates each base (Fig. 6.4). The peaks should be tall and well defined. If they are short and wide, it suggests that the sequence was "unclear" and you might see alternate bases or the letter N indicating that the nature of the base could not be determined. It is a good idea to sequence your insert in the forward and reverse direction so that you can check for these discrepancies.

Use a sequence alignment program such as EMBOSS (European Molecular Biology Open Software Suite) to compare the FASTA sequence to the original gene sequence (Fig. 6.5). If there appear to be different bases,

FIGURE 6.4 FASTA sequence.

```
Vector seq   TTTTGTTTAACTTTAAGAAGGAATTCAGGAGCCCTTCACCTATGCAATAC
                                                     ||||||||||
gene         ------------------------------------------atgcaatac

Vector seq   ACTCCAGATACTGCGTGGAAAATCACTGGCTTTTCCCGTGAAATCAGCCC
             ||||||||||||||||||||||||||||||||||||||||||||||||||
gene         actccagatactgcgtggaaaatcactggctttttcccgtgaaatcagccc

Vector seq   GGCATATCGCCAAAAACTGCTTTCTCTTGGCATGTTACCTGGCTCCTCTT
             ||||||||||||||||||||||||||||||||||||||||||||||||||
gene         ggcatatcgccaaaaactgctttctcttggcatgttacctggctcctctt

Vector seq   TTAATGTGGTGCGCGTCGCTCCACTCGGCGACCCCATTCATATCGAAACC
             ||||||||||||||||||||||||||||||||||||||||||||||||||
gene         ttaatgtggtgcgcgtcgctccactcggcgaccccattcatatcgaaacc

Vector seq   CGTCGTGTGAGCCTGGTATTACGCAAAAAAGATCTGGCCTTATTAGAAGT
             ||||||||||||||||||||||||||||||||||||||||||||||||||
gene         cgtcgtgtgagcctggtattacgcaaaaaagatctggccttattagaagt

Vector seq   GGAAGCGGTTTCCTGTAAG GGC GAG CTC AAT TCG AAG CTT GAA
             ||||||||||||||||| .|.
gene         ggaagcggtttcctgtgt taa------------------------------

Vector seq   GGT AAG CCT ATC CCT AAC CCT CTC CTC GGT CTC GAT TCT

Vector seq   ACG CGT ACC GGT CAT CAT CAC CAT CAC CAT TGA GTT TGA

Vector seq   TCC GGC TGC TAA CAAAGCCCGAAAGGAAGCTGAGTTGGCTGCTGCCA

Vector seq   CCGCTGAGCAATAACTAGCATAACCCCTTGGG
```

FIGURE 6.5 Comparison to the gene sequence.

check them against the trace to see if they are true substitutions or errors or problems with the sequencing read. In Fig. 6.5, there are no base changes, and the stop codon has been deliberately replaced in order to incorporate a tag at the 3′ end of the product (highlighted in red). Use this information to determine whether the construct is ok to use for subsequent downstream experiments.

SUMMARY

By now you should be familiar with the use of PCR to produce and screen plasmid constructs for proteins expression and analysis. This book will not provide additional information describing the preliminary experiments required to assess expression or protein purification because there are many comprehensive guides available for this. However, you should at least by this stage have a plasmid construct ready to use.

Chapter 7

Polymerase Chain Reaction for Knocking Out Genes

ABSTRACT

Polymerase chain reaction (PCR) can be utilized as a molecular tool for functional analysis of genes and can be used in concert with protein and gene expression to thoroughly describe the role of a given gene or genes. This chapter will describe how PCR can be used to prepare construct that are necessary to generate knockout mutations and how to assign function by complementing gene mutations using Kochs' molecular postulates. This chapter uses prokaryotes as a model for generating knockout mutations, and the procedures described will need to be adapted for single celled eukaryotes such as yeast and for more complex organisms, which are beyond the scope of this book.

7.1 INTRODUCTION

It is possible to analyze the expression of a gene, investigate its protein product and its function; the latter can be achieved by "knocking out" or inactivating the gene. The consequent phenotype gives an indication of the gene function, which can be verified by supplying the gene *in trans* on a plasmid in a process known as complementation, and which is often described as Koch's molecular postulates. The start point of these experiments requires identification of a target gene, as described in Chapters 1 and 2, but primer design is somewhat different. Consideration must be given to the type of gene knockout you which to achieve—do you want to disrupt the gene, replace it with an antibiotic resistance cassette or completely remove it? Each approach will be considered and described in this chapter. There are many other ways that knockout mutations can be generated; however, since this book focuses on polymerase chain reaction (PCR) only strategies that utilize this technique will be considered. This is not an exhaustive list of protocols but introduces the reader to two commonly utilized methodologies. The first relies on the use of a suicide vector, and the second uses the λ red recombinase system.

7.2 HOW TO KNOCKOUT A GENE USING PCR

A gene can be disrupted by the insertion of a foreign piece of DNA, often this is an antibiotic resistance cassette which allows for ease of screening. This chapter will focus only on knocking out genes by insertion of an antibiotic resistance cassette although there are many other strategies you can use—this methodology provides a fairly straightforward means of knocking out a gene and identifying mutant on the basis of antibiotic selection. Inserting an antibiotic resistance cassette into a gene can be done in a number of ways. One method requires PCR amplification of the gene of interest and the antibiotic resistance cassette. The gene sequence of commonly utilized antibiotic resistance cassettes can be easily found by searching gene databases, such as the NCBI database as described in previous chapters. Both genes are PCR amplified and cloned separately into plasmid vectors (as described in Chapter 6). A restriction site found within the gene of interest is identified, and during the PCR amplification of the antibiotic resistance cassette, this restriction site is incorporated at each end of the cassette (see Chapter 6). The plasmid containing the gene of interest is then linearized by digestion with the identified restriction enzyme, and the cassette is cut out of the second plasmid using the same enzyme (the fragment must be cleaned up by gel extraction as described in Chapter 6). The cassette and the linearized plasmid carrying the gene of interest are then joined by ligation. Subsequent screening is undertaken as described in Chapter 6. This method is fairly straightforward but disruption of genes in this manner can impact on the expression and function of downstream genes, meaning that you could be affecting more than just your target gene.

An alternative approach is to replace the entire target gene with the antibiotic resistance cassette. The following strategy is based upon:

> Taylor V.L., Titball R.W., Oyston P.C. (2005). Oral immunisation with a *dam* mutant of *Yersinia pseudotuberculosis* protects against plague. Microbiology 151, 1919—1926.

It is advised that you read this manuscript before proceeding with the experiments.

To achieve the above, you need to identify the entire target gene sequence including 500—1000 bp upstream and downstream of the gene. Primer design is a little more restrictive and requires two sets of primers. The first pair should amplify a fragment upstream of the gene and should include the start codon (upstream fragment); the second primer pair should amplify a region downstream of the gene, including the stop codon (Fig. 7.1).

You will need to include restriction sites that will be incorporated at the 5′ and 3′ end of the PCR products. Having PCR-amplified these fragments, they are each ligated into a T-tailed vector. These fragments are extracted using the restriction site incorporated at the 3′ and 5′ end of the fragments for

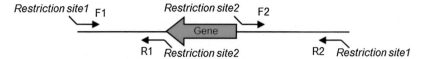

FIGURE 7.1 Identifying a target gene.

"upstream" and "downstream" regions, respectively. Both fragments are ligated into an appropriate vector cut with the same two enzymes (the vector you choose will be dependent on the organism you use). The resulting construct can be used to generate an unmarked knockout mutation or an antibiotic resistance cassette can be cloned into the gene for selection. For the latter, following propagation of the plasmid, it is linearized using "restriction enzyme" (RS2; Fig. 7.1) and ligated with the antibiotic resistance cassette (previously amplified and extracted from the initial cloning vector). As described in Chapter 6, this construct must be propagated and the insert verified as described.

A second alternative utilized inverse PCR. In this case, the entire target gene is initially PCR amplified and cloned into an appropriate vector (as described in Chapter 6). Next, primer pairs are designed to regions of the gene near to the start and end (Fig. 7.2).

These are used to amplify the entire plasmid plus gene fragments to produce a linear piece of DNA flanked by the start and end regions of the gene of interest. The primers utilized for inverse PCR have to be designed to incorporate restriction endonuclease sites that can be used allow the antibiotic resistance cassette (prepared as described previously) to be ligated into it, and also to allow the inserted to be "moved" to alternative vectors if required.

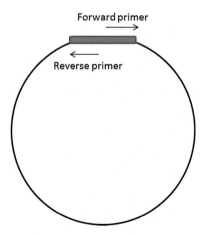

FIGURE 7.2 Target gene inserted into vector with forward and reverse primers.

Inverse PCR requires the use of special "long polymerases" that have proof-reading activity to ensure that the entire plasmid is accurately replicated. It is always necessary to sequence the entire construct after you have prepared it to ensure that the sequence remains unchanged. Base changes in the origin of replication, for example, could prove disastrous and of course you need to check that your gene fragment and cassette are also ok.

7.3 TRANSFERRING ENGINEERED PCR FRAGMENTS INTO THE TARGET ORGANISM USING A SUICIDE VECTOR

You will have constructed the "knocked out" gene using a "general" plasmid as described previously. In order to transfer it into your organism of choice, you will need to move it into a species specific plasmid, usually a suicide vector which cannot replicate in the recipient organism. It is advisable that you undertake a thorough search of the literature to identify an appropriate vector to use; most vectors will carry an antibiotic resistance cassette of their own or other means of selection. Once you have done this, you simply need to cut the fragment of DNA out of the plasmid you constructed it in and ligate it in. It is a good idea to have identified this vector early on so that the appropriate re-striction sites can be incorporated into the construct from the start.

Suicide vectors are conditional for their replication and are used to create duplications within the bacterial chromosome. When introduced into the recipient in which the plasmid cannot replicate, the entire plasmid is integrated into the chromosome. However, to replace a specific gene a double-crossover event is required; a small proportion of the population will have undergone a double crossover and you can differentiate between the two events by selection using both the cassette used to generate the gene knockout and the plasmid-specific selector.

Recipients that have undergone single-crossover events can be identified by their ability to grow under both selective pressures (the knockout gene and the plasmid); recipients that have undergone a double-crossover event can be identified by their ability to grow on media containing antibiotics specific for the cassette used to prepare the knockout construct. A simple way to do this is to select transformants and prepare replicate grid plates consisting of one or both selective pressures. Following incubation compare the replicate grid plates; recipients that have undergone a single crossover will grow on both plates, recipients that have undergone a double crossover will grow on the plate with antibiotics specific to the cassette, only (Fig. 7.3).

Unmarked knockout mutations can also be prepared in this way. For these knockout mutations, an antibiotic resistance cassette is not used to disrupt the gene of interest; instead the upstream and downstream fragments are ligated to produce a truncated gene product, which is subcloned into the suicide vector. This method is "cleaner" and is less likely to result in disruption to the

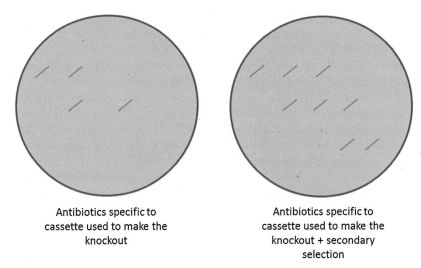

| Antibiotics specific to cassette used to make the knockout | Antibiotics specific to cassette used to make the knockout + secondary selection |

FIGURE 7.3 Comparing grid plates.

expression of other genes. However, it requires a means of counter selection to identify recipients in which double-crossover/knockout events have occurred.

To do this, suicide vectors carrying the *sac* gene derived from *Bacillus subtilis* are used. For gram-negative bacteria, expression of the *sac* gene is toxic in the presence of 5% sucrose which provides a means of selecting recipients that have lost the vector and therefore have undergone a double-crossover event. Recipients recovered on sucrose are subsequently screened for loss of antibiotic resistance (carried on the plasmid). Replicate grid plates as described previously can be prepared to achieve this type of selection.

For both marked and unmarked knockout mutations, it is important to screen the putative knockout mutants using PCR.

7.4 THE λ RED RECOMBINASE SYSTEM

This recombinase system relies on PCR but does not utilize any plasmid vectors. It is versatile and can be used with a wide range of prokaryotic microorganisms. Knockout mutation generated using this method relies on gene disruption by insertion of a kanamycin or chloramphenicol resistance cassette. Primers are designed to amplify the cassette, but must incorporate regions of sequence with homology to regions directly upstream and downstream of the gene of interest (30–40 bp incorporated in much the same way as a restriction site; see Chapter 6). PCR amplify the cassette; for clarity it can be ligated into a T-tailed vector at his point and sent for sequencing.

The recipient bacteria must possess the λ red recombinase system which is encoded on a plasmid (pKD46) and whose expression is induced in the presence of arabinose. You will therefore need to transform your recipient organisms (by electroporation). The plasmid carries the *bla* gene which encodes for ampicillin resistance and therefore can be maintained under ampicillin selection. The plasmid origin or replication is temperature sensitive, and it will be lost will be lost following a temperature shift from 30°C to 42°C.

You will need to prepare fresh electrocompetent cells for this procedure and prepare them in media containing ampicillin and 10-mM arabinose. All incubation steps must be carried out at 30°C (and so the cells might take longer to grow). Aside from the different growth media and temperature, the electrocompetent cells are prepared as previously described in this chapter.

Next, you need to electroporate the linear PCR-amplified fragment into the recipient, using the standard protocol described in this chapter. Allow the electroporated bacteria to recover at 37°C for 2−3 h. The absence of arabinose or ampicillin from the recovery media means that the recombinase system will be lost as the recombination event should have occurred. Plate the recipient organisms onto selective agar (choose the antibiotic based on the cassette you used) and screen for loss of the red recombinase plasmid by preparing replicate grid plates as previously described.

Putative knockout mutants can now be screened by PCR.

If you choose to utilize the λ red recombinase system, it is highly recommended that you read the two references given below to develop a greater understanding of the technique and its development:

Datsenko, K.A., Wanner, B.L. (2000). One-step inactivation of chromosomal genes in *Escherichia coli* K-12 using PCR products. Proc. Natl. Acad. Sci. 97, 6640−6645.

Poteete, A.R., Fenton, A.C. (2000). Genetic requirements of phage lambda red mediated gene replacement in *Escherichia coli* K-12. J. Bacteriol. 182, 2336−2340.

7.5 SCREENING FOR KNOCKOUT MUTATIONS

Firstly you will need to extract the chromosomal DNA from the presumptive knockout mutants; there are number of commercially available kits to do this or you can use the phenol:chloroform method outlined in Chapter 1.

Having extracted the chromosomal DNA, you will need to prepare a PCR reaction using the forward primer for your "upstream" gene fragment and the reverse primer for your "downstream" gene fragment. When used in this manner, the PCR product for the knockout mutant should be of a different size to that of the wild-type gene, i.e., it should be roughly the size of the antibiotic resistance cassette, or if you have generated an unmarked mutant, the size again should be smaller but you will need to work out the size based on the size of the gene region and the size of the fragment that was essentially "removed". If you generated a marked knockout mutation, you could set up at the same time, a second PCR reaction using the primers originally utilized to amplify the antibiotic resistance gene. In this case, you would only expect a PCR product for the knockout mutants. It is the best to use both strategies and not rely on the latter; using the latter alone relies on inferring a positive result from a negative reaction, which is not advisable.

SUMMARY

Having read and worked your way through this chapter, you should be able to plan and undertake the experiments necessary to produce a marked or unmarked knockout mutation in a prokaryotic microorganism. The strategies described here are not the only ones available and newer methodologies that have versatility spanning many domains of life are now known and utilized. One such methodology utilizes the CRISPR/Cas system (Clustered Regularly Interspaced Short Palindromic Repeats/Cas). Several research groups have pioneered protocols that utilize CRISPR/Cas and numerous manufacturers are beginning to develop "kits" that will be commercially available. It is recommended that you take the time to learn about this technique, and a literature search will yield a large number of published research articles describing CRISPR/Cas.

Appendix 1

Trouble Shooting

Problem	Cause	Possible Solution
General Polymerase Chain Reaction (PCR) Troubleshooting		
No chromosomal DNA or RNA following extraction	Insufficient sample	Use a larger sample to extract from
	Degradation	Check all reagents and equipment are DNAse/RNase free
Smeared chromosomal or RNA preparation on gel	Degradation	Check all reagents and equipment are DNAse/RNase free
Multiple PCR products on gel	Nonspecific primer binding	Check primer design
		Increase annealing temperature
		Decrease magnesium concentration
	Too many PCR cycles	Decrease number of PCR cycles
No PCR product on gel	Ingredient missing from reaction	Check reaction mixture
	Presence of PCR inhibitor	Add an additional cleaning step to your extraction method
	Too few PCR cycles	Increase number of cycles
	Extension or annealing time too short	Increase extension or annealing time
	Not enough template	Titrate template

Continued

81

—cont'd

Problem	Cause	Possible Solution
Smeared PCR product	Too many PCR cycles	Decrease number of PCR cycles
	Extension time too long	Decrease extension time
	Annealing temperature too low	Increase annealing temperature
	Concentration of polymerase is too high	Titrate polymerase
Primer dimmers	Primer pairs self-anneal	Check primer design
		Titrate primers
Product amplified from negative control	Contamination	Check work area is clean and "PCR-ready"
Low 260/280 ratios	Contamination or residual phenol in extract	Repeat extraction, including and extra chloroform/isoamyl alcohol (IAA) step
Quantitative PCR (Q-PCR)-Specific Troubleshooting		
Amplification occurs later than expected	Insufficient template	Titrate template
	Poor quality RNA	Repeat extraction
Sigmoidal amplification curve	Background noise/ fluorescence	Ensure samples are thoroughly mixed
		Check for formation of primer dimers that will bind the fluorophore
PCR efficiency falls below 90%	Inaccurate pipetting	Repeat the reaction
	PCR inhibitors	Clarify the quality of the extraction method and reagents
	Incorrect analysis of standard curve	Reassess the standard curve

Appendix 2

Suppliers

These are some representative examples of suppliers of PCR machines, reagent, and primers. It is by no means an exhaustive list and other suppliers are available and easy to find by internet search.

Reagents, consumables, and machines:
Biorad: www.bio-rad.com
Applied Biosciences: www.appliedbiosciences.com
Illumina: www.illumina.com
Life Technologies: www.lifetechnologies.com

Reagents (including primers and extraction kits)
Promega: www.promega.com
Sigma—Aldrich: www.sigmaalrdich.com
New England Biolabs: www.neb.com
Thermo Scientific: www.thermoscientific.com
Qiagen: www.qiagen.com
Bioline: www.bioline.com
Roche: www.roche.com
MWG: www.mwg-biotech.com

Index

Printed in the United States
By Bookmasters